Destination Anthropocene

CRITICAL ENVIRONMENTS: NATURE, SCIENCE, AND POLITICS

Edited by Julie Guthman, Jake Kosek, and Rebecca Lave

The Critical Environments series publishes books that explore the political forms of life and the ecologies that emerge from histories of capitalism, militarism, racism, colonialism, and more.

1. *Flame and Fortune in the American West: Urban Development, Environmental Change, and the Great Oakland Hills Fire,* by Gregory L. Simon
2. *Germ Wars: The Politics of Microbes and America's Landscape of Fear,* by Melanie Armstrong
3. *Coral Whisperers: Scientists on the Brink,* by Irus Braverman
4. *Life without Lead: Contamination, Crisis, and Hope in Uruguay,* by Daniel Renfrew
5. *Unsettled Waters: Rights, Law, and Identity in the American West,* by Eric P. Perramond
6. *Wilted: Pathogens, Chemicals, and the Fragile Future of the Strawberry Industry,* by Julie Guthman
7. *Destination Anthropocene: Science and Tourism in The Bahamas,* by Amelia Moore

Destination Anthropocene

Science and Tourism in The Bahamas

Amelia Moore

UNIVERSITY OF CALIFORNIA PRESS

University of California Press
Oakland, California

Library of Congress Cataloging-in-Publication Data

Names: Moore, Amelia, 1981- author.
Title: Destination Anthropocene : science and tourism in the Bahamas / Amelia Moore.
Description: Oakland, California : University of California Press, [2019] | Series: Critical environments: nature, science, and politics ; 7 | Includes bibliographical references and index. |
Identifiers: LCCN 2019000631 (print) | LCCN 2019002842 (ebook) | ISBN 9780520970885 (e-edition) | ISBN 9780520298927 (cloth) | ISBN 9780520298934 (pbk.)
Subjects: LCSH: Climatic changes—Effect of human beings on—Bahamas. | Biocomplexity—Bahamas. | Tourism—Environmental aspects—Bahamas.
Classification: LCC QC903.2.B2 (ebook) | LCC QC903.2.B2 M66 2019 (print) | DDC 304.2097296—dc23
LC record available at https://lccn.loc.gov/2019000631

Manufactured in the United States of America

28 27 26 25 24 23 22 21 20
10 9 8 7 6 5 4 3 2

For Eleanor and Emma

Contents

Illustrations

MAP

FIGURES

Acknowledgments

This book originated as a strange dissertation, and so I would like to begin by thanking the members of the Anthropology Department at UC Berkeley, especially Cori Hayden, Paul Rabinow, and Donald Moore, for their assistance with many of the foundational ideas in this book. In addition, I must thank Charis Thompson in the Department of Gender and Women's Studies for her welcome advice. My fellow graduate students also deserve my thanks for their heaps of support over the years, especially Alfred Montoya, Shana Harris, Beatriz Reyes-Foster, James Battle, and Theresa Macphail. Thank you for the phone calls, the emails, the texts, the Facebook groups, and the conversations over many cups of tea and pints of beer. This book is your book too.

The fieldwork that went into this book took place over several years in The Bahamas, and I must thank the fisheries officers at the Department of Marine Resources in the Ministry of Agriculture and Marine Resources for issuing me research permits and for participating in my research. I owe a debt to the University of The Bahamas for sponsoring my work, especially to Linda Davis, Jessica Minnis, Nicolette Bethel, John Cox, Stephen Aranha, Brenda Cleare, Michael Stevenson, Lisa Benjamin, Pandora Johnson, and Margo Blackwell. The Bahamas Reef Environment Educational Foundation provided me with insight and access to its office, and the Bahamas National Trust, Nature Conservancy of The Bahamas, Ministry of Tourism, Ministry of Environment, Antiquities Monuments and Museums Corporation, Department of Archives, and National Public

Library were all essential organizations whose staff did me many favors. I am also deeply indebted to Margot Bethel and the late Hub Community Arts Centre. Additionally, I must mention my dear friends Kareem Mortimer, Jonathan Morris, Isabella Boddy, Monique Wszolek, Andrew Jones, Alex Wassitsch, Augusta Wellington, Casuarina McKinney, Mallory Raphael, Eddie Raphael, Michael Edwards, and Adelle Thomas for their unfailing advice and enthusiasm, and I must especially thank Bryan Boddy for his invaluable commentary on several drafts.

Paige West, at Columbia University's Barnard College, without whom I would not have entered graduate school, changed the course of my life and must be thanked, even though she cannot ever be thanked enough. I must thank Fiorella Cotrina for believing in me, as well as everyone at the Abess Center for Ecosystem Science and Policy at the University of Miami, including Gina Maranto, Andee Holzman, and most especially Kenny Broad, who brought me to The Bahamas in the first place and who continues to inspire me with research possibilities. Laura Ogden, Danielle DiNovelli-Lang, Julie Guthman, and Kate Marshall provided essential reviews and edits and should be thanked by all my readers. Any remaining problems with the text are entirely of my own making.

I am appreciative of the faculty in the Department of Marine Affairs at the University of Rhode Island for their patience with my evolving publication schedule. I cannot forget to thank Scott Andrews, who helped me through the longest stretch of my fieldwork. Finally, I must thank my parents, Sandra Walker and Wesley Moore, and my brother, William Moore, for their love, advice, and support in all things.

The research for this book was funded at various times by the American Museum of Natural History, the National Science Foundation's Graduate Research Fellowship Program, the Abess Center for Ecosystem Science and Policy at the University of Miami, the Wenner-Gren Foundation for Anthropological Research, and the U.S. Fulbright Scholar Program.

Abbreviations

AMMC	Antiquities Monuments and Museums Corporation
DMR	Department of Marine Resources
GCS	Global change science
IPCC	Intergovernmental Panel on Climate Change
MPA	Marine protected area
NSF	National Science Foundation
SIDS	Small Island Developing States

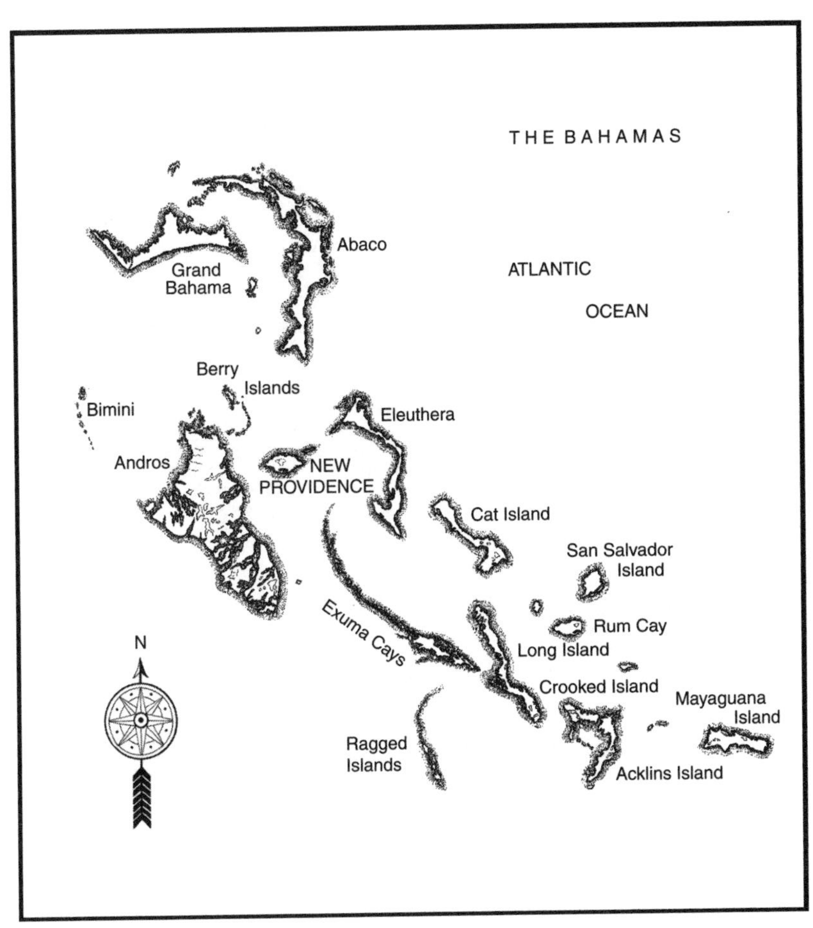

Map of the major islands of the northern and central Bahamas. Image created by Quail Lane (2017).

Introduction

The Anthropocene Islands

ISLAND MAKING

> Arriving at destination Anthropocene entails the prospect of facing profound changes in the global political, economic, environmental, and social order, along with a wide range of possible consequences.
>
> —Martin Gren and Edward H. Huijbens, *Tourism and the Anthropocene*

Islands are rarely what they seem. I am acquainted with a small set of islands that have been visited by millions, and yet most visitors know very little about them. In this set of islands, the land and ocean are one continuous flow of energy. The substance of the land consists of exposed dunes and shallow banks formed by the precipitation of calcium carbonate from seawater. This limestone was excreted from the sea through the bodies of tiny sea creatures and shaped by chemical reactions repeating over millennia. Emerging on the western edge of the subtropical Atlantic Ocean, these islands are home to a host of life forms across the land and sea world: a kaleidoscope of reptiles, birds, insects, corals, fish, and shellfish that live in pine barrens, tropical coppice, mangrove marls, undersea reef colonies, sand flats, and beds of sea grasses. A shifting human populace of citizens, migrants, and international tourists also reside in this archipelago. These are the seven hundred islands of The Bahamas.

The islands of The Bahamas are icons of tropical paradise, internationally known for their white-and-pink sand beaches, hedonistic resorts, and aquamarine seas. But this island imaginary is in transition.

FIGURE 1. The Hub Community Arts Centre, Nassau. Photo by author (2012).

For example, take recent events in cave diving. On some islands the sea not only surrounds the land but flows within the porous limestone rock of the islands themselves, creating fantastical underground and underwater caverns. This world was recently explored by visiting scientists whose submarine journey was captured in an issue of *National Geographic Magazine*. The striking cover photo depicts two vulnerable divers in black wetsuits hesitantly swimming in a mysterious submerged tunnel. They are surrounded on all sides by jagged, glimmering rock formations bathed in a blue glow that is simultaneously ethereal and eerie. It almost looks as if the explorer scientists are shining their flashlights into the toothy mouth of a giant island monster poised to swallow them whole. Far beyond the famous beaches, this cave system, or "sea within," is one of the archipelago's most wondrous and underappreciated geologic features—for the time being.

Unbeknownst to most people, the Bahama Islands have become a destination for earth scientists of many stripes, united in their mission to reveal the planet's secrets. In this instance aquatic caves are valued by scientists because they contain records of climatic change. They also contain artifacts of the human occupancy of the islands left by several waves of indigenous populations. Some of these artifacts have been linked to the ecological impacts of that occupancy. And fossils have been discovered in the caves that have been used to demonstrate the evolution and extinction of a number of endemic species spanning across millennia. Thus, these cave systems are, for earth scientists, time capsules that contain evidence of the long-term effects of anthropogenesis on the Bahama Islands and the world.

On an even more recent warm spring day, safely above ground, I ventured in a van down old logging roads in the pine forests of the island of Abaco, with Marcus, a U.S. entrepreneur. The van was full of dive gear. As we bumped along through arid woods, he explained that the forest, the road, and in fact the entire landscape of this part of the island concealed a whole universe beneath the ground. There was water all around us, he insisted, flowing through myriad tunnels and caverns that we could not see but that were right below the forest floor. It was his dream to expose these waterways beyond photography, guiding visitors through an experience with submarine space and geologic time.

Exposure, for Marcus, would come two ways. First, as a professional cave diver, he would lead clients into the same tunnels and caverns where scientists had recently found evidence of ancient life in the islands. But he knew this experience was limited to the few visitors with the certifications and wealth to access such an adventure. This realization sparked his second plan to open the subterranean world to a more general public by bringing it above ground. In addition to leading dive tours, he wanted to create more accessible terrestrial pathways through the forest. These paths would look just like typical hiking trails, except that these trails would mirror the winding tunnels beneath, and interpretive signs with dramatic images of the caverns and artifacts below would be posted at corresponding sites along the way. This strategy would allow the underground universe to be traversed on foot between the trees. His plan would make the mysterious geologic features real in a park setting for visitors to experience the caves both above and below the ground.

As an entrepreneur, Marcus wants to help tourists "discover" the island's underground universe and its record of human and ecological change. He hopes that visitors will pay fees to help keep this valuable island feature safe from more destructive forms of resort development. However, as the *National Geographic* cover story shows, these tunnels and caves have been in the process of being "discovered" by international scientists for several years. In fact, Marcus's first encounter with these underground features was as a research diver, assisting in cave exploration, and he himself was one of the divers featured prominently in the magazine's cover article.

These submerged caves are already known to Bahamians as "blue holes." On rural islands like Abaco, Andros, Grand Bahama, or Eleuthera, Bahamians have long-standing relationships to these geologic features, forged over generations through hunting, fishing, bathing, and even drowning in the holes. In one Bahamian settlement there is even a blue hole smack in the center of town, forming a park. Children joyfully play in this hole on summer Sundays, leaping from the limestone edge to create plumes of water, forever trying and forever failing to swim to the bottom. But such vital relationships have weakened in recent years, as islander livelihoods become more dependent on wage labor, tourism, and foreign investment and less reliant on island ecologies.

Here then, is the paradox that surrounds this entire book: Marcus's plans are one example of science and tourism entangling together as they create new businesses, brands, and destinations for a changing planet, but their solution to the problem of anthropogenesis is more exposure for those parts of the world that are already the most susceptible to ecological change and economic dependency.

. . .

Of course, The Bahamas has a long history with tourism and thus experience with exposure. This history has not been static. The way tourism officials position the islands has evolved over time to make The Bahamas more easily consumable for the outside world of potential investors and visitors. This consumption has been facilitated through the promotion of the islands as a timeless tropical sanctuary from the cares of daily life in the industrialized world. This is what I think of as the "Isles of June" discursive framing of island place, a phrase attributed to Christopher Columbus that was used to describe the region as a tourist destination during the 1940s, when the islands were still governed as a segregated British colony.[1] Within this frame the romance of colonialism is maintained through beach resorts for foreign "guests" served by native "hosts" in a carefully orchestrated illusion of perpetual tranquility.[2]

In the Isles of June, the (white) guest is always right, the (black) native server is always cheerful, the weather is always warm, and there is no reason to worry about histories of racial subjugation, corrupt local politics, or the uneven impacts of tourism.[3] In the Isles of June, foreign investors build large hotel complexes on coastal property, clearing the bush and cleaning the beaches of flotsam. Tourists arrive in gawking throngs, disembarking from enormous cruise ships or jumbo jets. These arrivals vacate cold and mundane lives at home to sun themselves by day and eat and drink at hotel restaurants by night, exploring the surrounds by tour bus or party boat, if they explore at all. In the Isles of June, many Bahamian citizens aspire to "good hotel jobs," and the tourism sector is promoted by government as the engine of local employment. Bahamians live to serve.

This "peaceful" discourse of place, dominant throughout the majority of the twentieth century, was complicated in the 1970s by the advent of an environmental consciousness led by U.S. ecologists. This discourse took shape alongside Bahamian independence from Britain as scientists reframed the archipelago as the "Ephemeral Islands" of wild species and fragile ecosystems subject to the strains of local environmental degradation and extinction.[4] The Ephemeral Islands are the domain of naturalists and environmental educators who are intensely focused on maintaining the diversity of island species and informing the public about their "natural heritage." These scientists come from foreign universities to conduct site-specific research on subjects such as island-adapted lizards, grouper spawning aggregations, or ground-nesting parrots. In the Ephemeral Islands Bahamian citizens are assumed to be largely unaware of the natural wonders that surround them.

The Ephemeral Islands frame is now quite familiar, and it uneasily coexists with the Isles of June. In fact, we don't have to go to The Bahamas to be familiar with narratives like these or to know that environmental advocates, educators, and scientists are frequently at odds with the tourist industry over development, habitat loss, and pollution. More recently, I have discovered another framework stemming from the discursive and practical entanglement of science and tourism, which I call the "Anthropocene Islands." In this frame local environmental concerns are described by researchers and entrepreneurs as evidence of ongoing global environmental change. Rather than perpetually clashing with the Isles of June, the Anthropocene Islands orientation appears to be radically inclusive of local communities and economies, providing opportunities for the tourist industry to reinvent itself as part of a global market for sustainable tourism destinations. This book interrogates the assumptions of the Anthropocene Islands.

Islands are not inert geologic objects. Island are both created and creative. This becomes apparent by following the redevelopment of particular islands across space and time. In the years spent conducting the research for this book among members of the tourism industry, scientific researchers, and island residents, I learned that the Bahama Islands aren't just evolving geologically, accreting and eroding over millennia; they are changing at the speed of thought. The Bahamas is not alone in this process of re-creation. Some islands—small islands—are now made, in part, within the crucible of the Anthropocene idea and its logics and technologies. The Bahamas is a small place, not commonly associated with major world events, but I have found it to be just the right kind of place in which to examine the Anthropocene idea in action.

* * *

"The Anthropocene" was coined within the earth sciences. It labels the period of recent planetary time in which human activities began to influence all of the planet's biological, geologic, and chemical systems. Most scholarship about the Anthropocene defines it as a problem that should engage our attention.[5] Many scholars look to assign responsibility for the problem, accurately chart its evolution, and propose solutions. This represents a sea change in the environmental movement, transforming the "environment" into an "epoch." But this transformation should not be taken at face value. The Anthropocene can also be understood as evidence of a philosophical shift in the way many people come to know their world, themselves, and their relationship to other living things.

This book is about the Anthropocene idea itself as a response to perceived planetary change. This idea inspires new entanglements of industries we normally think of as distinct and remakes the world in ways that benefit certain international industries at the expense of more locally engaged ways of life. Emergent collaborations between the sciences of global change and tourism are changing not only markets for travel but the very spaces and places where we live, work, and vacation. It is critical to understand how this is happening, what these emergent collaborations are making, and what that means for specific places and people in the world. For example, island places like The Bahamas were historically used in the colonial period as laboratory experiments for science and commerce. This book shows how the Bahama Islands of the twenty-first century are once again on the front lines of planetary change and new designs for living that may one day affect everyone.

Anthropology, as the study of humanity across space and time, is uniquely capable of examining transformations of knowledge about life. Anthropologists are well positioned to document the Anthropocene as an emergent way of understanding and experiencing the world around and within us. This approach to the Anthropocene, as a powerful idea with the potential to remake human relationships to the world, is a needed counterpoint to the proliferation of media about global change slowly making its way from academia to various publics. This anthropological approach can and should take many forms.

The approach I take in this book is most readily referred to as Anthropocene anthropology. My version of Anthropocene anthropology centers around processes of global change, including but not limited to climate change, as they are happening. My work focuses on the ideas and industries that remake the world in response to Anthropocene concerns, because responses to change are themselves a large part of the overall process of global environmental change. Anthropocene anthropologists are not dispassionate observers but instead are frequently involved in the proposed solutions and problems they study.

I am a sociocultural anthropologist belonging to a long tradition of scholars who study the consequences of transnational ideas about the environment. I examine the events these ideas inspire from both inside and outside the networks that create them. I am also American, born with no immediate ties to The Bahamas or the Caribbean. My relationship with the Anthropocene Islands began years ago as I first became a global change scientist and later an observer of global change science itself. My anthropological attention has led me to know a number of scientists, the sites of their projects, and the policy circles and travel markets that institutionalize their work. I came to know these people and the relationships they helped build because I was intimately involved in their research as a student, volunteer, and participant observer. Over time I developed a sense of what these researchers do as a loose group of traveling actors.

This book is based on fifteen months spent living in The Bahamas, utilizing research methods that included ethnographic interviews and participant observation from 2007 to 2009. Shorter trips to the country informed this research in 2002, 2005, and 2006, along with multiple annual supplemental visits between 2010 and 2016. The five main chapters of this book are written through the lens of my own island-hopping experience with fieldwork. These chapters ground the book by presenting different examples of global environmental change produced at the intersection of science and tourism in the islands of The Bahamas during this fourteen-year period.

My strategy has been to involve myself with projects on different islands in The Bahamas to observe the Anthropocene idea in action. I was a student field assistant collecting social surveys about fishing in a small village in Abaco. In South Eleuthera I developed a relationship with an island-based boarding school for visiting U.S. high school students. I volunteered for environmental nongovernmental organizations in New Providence. The examples that compose this book are clustered among the central and northern Bahama Islands, where the majority of the population resides, the majority of tourism occurs, and most field science is carried out. The method leads to the product: I provide a collection of ideas, moments, places, and stories that come together to show how islands are being remade. Because there is no center anchoring all these events, I become the center of the story in this telling. I write from my perspective as a global change science practitioner but also as a second order observer of global change science.[6] This is the best way I have found to represent the condition of living and working in the archipelago.

In the following pages I outline the core concepts that define my understanding of the Anthropocene Islands. First, I explore the history of the Anthropocene concept, showing how it has become increasingly significant for natural scientists, social scientists, and humanists as a powerful creative idea. Second, I describe global change science as the industry responsible for reproducing and legitimizing the Anthropocene idea around the world. Third, I explain why islands are iconic spaces in the Anthropocene and an important vantage from which to study the symbolic and material effects of the idea. The chapter culminates in the conjuncture of tourism and field science as the engine that is remaking the postcolonial Bahamas in the name of anthropogenic planetary change. The outcome is one of the first ethnographies of the Anthropocene that shows how a specific locale is being remade into an emergent laboratory for the production of scientific knowledge about the environment, tourism as a commercial enterprise, and collective understandings of contemporary life.

CONCEPTUALIZING THE ANTHROPOCENE

> Anthropocene: the period of time during which human activities have had an environmental impact on the Earth regarded as constituting a distinct geological age.
>
> —*Merriam-Webster*

The Anthropocene is about many things. For instance, it is often associated with climate change, but it is about even more than that. We can see this clearly in The Bahamas. Once we look past the manicured hotel beaches and imported palm tree–scapes that dominate most tourist destinations in the archipelago, evidence of more anthropogenic environmental change is everywhere.[7] Along the coasts beaches erode into the sea as a result of coastal development that has removed protective dunes and mangroves. Long-term residents of Nassau recall that the sandy shores of Cabbage Beach or Cable Beach used to be far bigger when they were younger. The sea itself is also changing. Fishers tell of the oceans of crawfish that used to crawl across the banks, the acres of elkhorn coral they dove among as children, and the gargantuan grouper aggregations their fathers once fished. Many agree that these sights are gone as a result of overconsumptive fishing, while they notice a new arrival in near-shore waters: the invasive Indo-Pacific lionfish. And on the island of New Providence the last acres of green space are enclosed in parks, preserving the memory of childhoods spent "in the bush," as neighborhood subdivisions and tourist enclaves appear to cover every available surface.

These Bahamian changes are distinct from one another, occurring in different locales and over different timescales. Yet, for some scientists, these changes are linked by the idea that they are all the result of human actions replicated all over the world, constituting a global event. From the immediacy of fishing to the effects of greenhouse gas emissions occurring a world away, local environmental change is tied to the idea of anthropogenesis as a global phenomenon. This global phenomenon—this event—is now encapsulated by a geologic age that some earth scientists propose to call the Anthropocene.

There is a long history in the sciences of attempting to diagnose the unique condition of the present in a way that differentiates the present from all that has come before.[8] The Anthropocene is an attempt to do just that. But like other attempts at diagnosis, any definition of the Anthropocene says as much about its makers as it does about the present.[9] Any conceptual analysis of the Anthropocene must explore the term's affective qualities, muddled politics, and tensions that have the capacity to reshape material worlds.[10] In short, while the concept seems clear on its face, the Anthropocene has multiple, contested valences.

The term originated with natural scientists, presented in 2000 by the chemist Paul Crutzen and the biologist Eugene Stoermer to describe the impact of human life on Earth. The term is intended to equate human activities with earth systems like the geologic carbon cycle, putting a technical name to notions of anthropogenesis that have animated debates about the environment for decades.[11] The proposed term labels the planetary present as an unstable geologic epoch following the formally recognized Holocene, a period of interglacial planetary stability. Many scientists argue that the Holocene has been displaced by the Anthropocene as a direct result of human activities that influence global processes to an unprecedented extent.[12] These activities stretch beyond the industrial production of greenhouse gases alone.[13] They include population growth, globalized economies, the industrial use of natural resources and materials, industrial agriculture, systematic overfishing, freshwater consumption, land surface use, habitat destruction, and deforestation.[14]

The Anthropocene idea has gained popularity in the scientific community because it presents an opportunity to study and manage the many faces of global change, uniting seemingly divergent themes.[15] For example, as a global event stemming from human action, the hole in the ozone layer can now be categorized with events such as altered nitrogen and phosphorus cycles, biodiversity loss, fisheries decline, pervasive invasive species, plastic pollution, and the many possible effects of climate change. The idea is also popular among its advocates because of its supposed public appeal and accessibility; it is one relatively simple term signifying the complex political battle to acknowledge the human impact on the planet's systems.[16]

For the natural sciences the Anthropocene is a useful term for labeling a crisis situation. Within this logic humans are a geologic force, driving unprecedented planetary change. Beyond climate change and the advent of the sixth mass-extinction of life on Earth, scientists have determined that domesticated animals are now the majority of living vertebrates, and some speculate that human activities even influence the movement of tectonic plates.[17] But there is debate about whether the Anthropocene signifies the total destruction of life as we know it or if it represents a more inclusive sense of what counts as "natural."[18] Is the goal of the Anthropocene idea to inspire the protection of natural systems that remain the least influenced by human activity?[19] Or does it signal the final abandonment of the modern concept of pristine, natural systems removed from human influence in favor of managing multinatural, working landscapes?[20] These divergent interpretations somewhat anxiously coexist for natural scientists in the present moment.[21]

The term *Anthropocene* has not yet been scientifically consecrated to officially replace the Holocene at the time of this writing. But whether or not it is ever institutionalized as a formal geologic period, the Anthropocene is a powerful term that merges aspects of human and natural history to argue for their interdependence.[22] It is this sense of the

Anthropocene, as an *idea* that represents an emergent understanding of life itself, that should concern social scientists as much as its formal adoption or the everyday use of the term in practice.[23]

For social scientists the Anthropocene is perhaps the most sweeping idea stemming from contemporary science about the pervasive effects of human activities across space and time. But because of its expansive scope, the possibilities for intervention in the Anthropocene are quite murky.[24] Considering its potential power and the fact that it can be understood by naturalists in different ways, the Anthropocene idea requires further inquiry as we follow its evolution as a social meaning-making concept. This work has already begun.[25] For example, some historians have concluded that the human can no longer be separated from the natural in humanistic thought.[26] Humanist scholars are then challenged to describe the emergent figure of the Anthropocene human and to craft new methodological tools in the social sciences and humanities to mark these changes.

I have chosen to think of the Anthropocene idea as a creative force in itself, recognizing that concepts like this don't take shape in a vacuum. Even universalizing ideas evolve depending on the social and material terrain in which they are deployed.[27] Investigating the Anthropocene requires grounded specificity and an awareness of the scientific research, forms of capital accumulation, and social politics that help materialize the idea in specific locales. The terrain of The Bahamas is shaped as much by Caribbean colonial history and the postcolonial dependence on tourism as by the weathering of wind and water on limestone. The Anthropocene idea grapples with this terrain as it becomes an emergent reality for social life in the Caribbean and its small islands, framed by the travel industry, the science of global change, and islanders themselves.

For contemporary Caribbean anthropologists, regional research problems have included a focus on evolving forms of social difference in the form of race, class, gender, sexuality, and national belonging and identity. To this list I would now add the problem of anthropogenic impacts on island environments. These are the research problems within which the means for questioning contemporary life are possible; these are the problems that deserve anthropological attention.[28]

As an Anthropocene anthropologist working at the intersection of the natural sciences, social sciences, and humanities, I recognize that local ecologies, planetary processes, and human natures are mediated by multiple scientific, commercial, cultural, and political institutions. Therefore, the chapters of this book must pull together multiple responses to perceptions of the Anthropocene. These responses become pieces of an anthropological puzzle.[29]

When conceptualizing the Anthropocene as a kind of puzzle that comes together to reveal the way The Bahamas is currently being remade, it is important to ask what kind of Anthropocene idea is being narrated and by whom, which narratives have the most power, and what those narratives can make.[30] These questions demand that we observe those who turn narratives into material reality. It means conducting research among experts and entrepreneurs in places where both "problems for humanity" and their solutions are conceived and created. This mode of inquiry suggests that life in the Anthropocene is constantly undergoing a creative design process toward a continually shifting future.

GLOBAL CHANGE SCIENCE AND THE ANTHROPOCENE

Offshore of New Providence, a young Bahamian graduate student expertly jumps overboard in her dive gear. Because I have chosen not to dive, I am left on the research vessel to chat with the Marine Resources officer at the helm. It is another hot day in an endless stream of hot days, and it is refreshing to be at sea with the wind in our faces. The student has redesigned the seafloor below into an experimental substrate that mimics near-shore debris fields to track the rate of lionfish invasion in junk-polluted waters. She will soon have a master's degree and start her doctoral work at a Canadian university. As the boat sways in the waves, the officer and I talk quietly about the way the student's research might affect future marine policy in The Bahamas.

I find myself outdoors after dark walking around the grounds of a small-island research station in Eleuthera. I left my room in the new quarters for visiting researchers because I can hear too much through the vents designed to "naturally" move air around the structure without the use of air conditioning. I could hear another U.S. scientist negotiating with his partner at home about how long he will be abroad, and listening in made me uncomfortable, having had too many of those conversations myself. Outside it is pitch black, but I can hear the tide coming in, the call of night birds across the mangroves, and the soft whirring of the station's wind turbine.

Another hot day of waiting above the surface of the water—only this water fills a terrestrial cave deep in the bush on the island of Abaco. I am waiting for the cave-diving team to resurface with their film gear and extra tanks. They are U.S. scientists making a documentary about the secrets of these aquatic caves. Because I do not dive, and because I especially do not cave dive, I am stuck talking to the production assistant. We swat mosquitos and swap stories. He has worked with teams like this before, all over the region, checking equipment for science divers and explorers who can reveal underwater worlds many of us will never see in person.

It is freezing cold inside the ballroom of the Wyndham Hotel on Cable Beach, Nassau, and I shiver in my business casual attire. I am finally in one room with representatives of all the major environmental nongovernmental organizations and government ministries in The Bahamas at a regional Climate Change and Tourism conference. After listening to the jovial greetings exchanged between men and women in the audience, including many who I know to be professionally at odds over policy and practice, it strikes me that they all know one another well after years of working across the country and the region. The lights dim and an invited consultant from the United Kingdom, whom no one here has met before, begins to speak from the podium.

All I want to do is stay inside and hide. I have become dependent on the limited comfort of a small rented cottage in the tiny community of Cherokee Sound, Abaco, but I am supposed to be outside administering surveys to reluctant townspeople. I know they do not like completing these surveys about their use of the marine environment and their fishing practices. By extension, I assume they do not like me either. I feel young and foolish. What right do I have to ask these people to talk to me? But my research supervisor, an accomplished field-worker, will have none of it. "Stop lazing around the house!" he insists. "Get out there and make your quota! Only the dogs bite."

These vignettes trace the contours of an archipelago perpetually remade by science and scientists. The Anthropocene idea and thus the concept of the Anthropocene Islands would not exist if it weren't for the rise of global change science. Global change science (GCS) is a label I apply to a broad range of science that focuses on anthropogenic change in earth systems across scales. The transnational research practices of GCS scientists reproduce the idea of the Anthropocene whether or not they explicitly use the term itself. GCS is characterized by a research interest in the systemic relationship between humans and their environments and the socioeconomic drivers of environmental change.[31] I have worked with global change scientists throughout my academic life. They have been my professors, team members on field research projects, fellow students, mentors, colleagues, research subjects, and friends. These men and women are ecologists, biologists, behaviorists, and social scientists—including fellow anthropologists—interested in environmental change. They primarily work within universities, but many work for governmental agencies or nongovernmental agencies.[32] While these professionals come from distinct fields, they share a commitment to the interdisciplinary study of anthropogenesis and the belief that their research should inform policy to maintain biodiversity and stabilize planetary cycles of species, water, carbon, and nutrients. In The Bahamas, as in other small postcolonial nations, these professionals tend to be foreign, living and working in Europe, the United States, or Canada. They frequently travel to conduct research in ecologically vulnerable areas around the world, often with teams of other global change scientists.

There are a number of organizations that expressly focus on global change research and policy, the most famous being the Intergovernmental Panel on Climate Change, but there are smaller organizations of GCS scientists that treat global change as more than climate change and that study global change as a highly localized phenomenon. Organizations such as the Nature Conservancy, the Bahamas National Trust, the Bahamas Reef Environment Educational Foundation, the Perry Institute for Marine Science, the University of The Bahamas, and several U.S. universities and colleges are all partly responsible for the extension and perpetuation of GCS research and concerns in the Bahama Islands. This occurs through funding, recruitment, networking, and the deployment of scientists to conduct research on environmental change.

One of the most prevalent traits of transnational GCS projects is interdisciplinary collaboration. The perceived need for interdisciplinarity stems from the argument that global change is a complex, multifaceted process that cannot be encompassed by any one discipline. Interdisciplinarity can take many forms. For example, research projects often combine different quantitative fields within the natural sciences.[33] The Anthropocene idea, however, with its orientation toward the impacts of human activities, increasingly promotes collaborations between the natural and social sciences. The social scientists that are often included in global change research tend to be those whose methods can most easily be integrated into natural science research, including quantitative economists, sociologists, and behaviorists.[34]

The style of scientific thought operating within such interdisciplinary Anthropocene research is often referred to as socioecology. Socioecology in this case refers to the study of human populations and their systemic interactions with nonhuman processes.[35] GCS therefore utilizes a form of reason that I call *socioecologics,* in which the human and the natural are coupled systems that must be understood holistically in management planning.[36]

There are tensions between forms of socioecologic within GCS research projects. For many projects one path toward understanding human and environmental relationships is the mathematical modeling of complex systems.[37] This inherently favors quantitative researchers whose data can be integrated with data from other disciplines to form the models that define current socioecological relationships and predict the effects of future change. But social researchers whose work is qualitative in nature and whose analysis is interpretive are often incompatible with the modeling concentration of much GCS research. Therefore, GCS research is self-consciously interdisciplinary, but this interdisciplinarity is limited in its form and content. The key is that the socioecologics deployed in research determine the kinds of sociality and living systems that are recognized within GCS projects.[38] These projects then produce knowledge based on their assumptions about sociality and life, knowledge designed to inform the governance of socioecological systems.

In response to GCS research delineating the contours of the Anthropocene, the governance of human and nonhuman life is slowly evolving. The governance of human bodies and populations (what some scholars call biopolitics) and the management of nonhuman ecosystems (what other scholars call ecopolitics) are no longer considered distinct but have in fact been merging for decades.[39] GCS research self-consciously links global environmental change and nonhuman life to human health and well-being for populations, individuals, and the human species. Thus, the idea of the human is changing with the Anthropocene idea, becoming inseparable from nonhuman organisms and the environment. A more apt concept to describe this current merger within Anthropocene governance might simply be "ecobiopolitics."[40]

Therefore, for many scientists, the human is now best thought of as a socioecological process itself, becoming a prime component of Anthropocene systems. Just as humanist scholars are recognizing that they cannot study humanity in a vacuum without acknowledging the role of the natural world in human lives, GCS scientists increasingly claim there are no ecological systems that can be understood without a central focus on human actions as drivers of change. The new object of calculation is not the social or the natural but rather a combination that makes a kind of whole. For some scientists planetary systems and human activities are now understood sociecologically and governed with integrative management techniques at multiple scales.[41]

As GCS slowly changes common understandings of the interrelation between the human and the nonhuman, socioecologics are becoming key components of the Anthropocene idea and its capacity for re-creation. In The Bahamas the advent of GCS socioecologic is reshaping the islands of the archipelago into sites for the discovery of vital knowledge about a changing planet. My experiences with GCS research reveal how far-reaching socioecologics can be, how they influence place through ecobiopolitical policies, and how they privilege certain knowledge and methods over others. The stakes are high because as GCS projects and policies remake places in the name of the Anthropocene, they also reformulate identities for both scientists and laypeople.

SMALL ISLANDS IN THE ANTHROPOCENE

> It is often said that the world's small island states are on the frontlines of climate change. Many are isolated, low-lying, and sit in the path of powerful tropical storms. Others face seasonal and, at times, severe droughts. . . . But, about three decades ago, some began to notice that these weather extremes were becoming more and more intense. . . . It was becoming increasingly clear that the world faced a serious environmental challenge and that islands were particularly vulnerable. Something had to be done.
>
> —Alliance of Small Island States

This excerpt is an attempt by the Alliance of Small Island States to make their important activism visible within the institutional history of international climate change negotiation. The "Anthropocenic" space that enables this move, and indeed the alliance itself, is the small island. This is one example of the way that the Anthropocene idea, enabled by scientific exploration and experimentation, has begun to reproduce spatial formations.

Islands have a special foothold in the Western scientific imaginary as places of significant biological and social processes. For instance, Charles Darwin's research in the Galapagos is a canonical example of the role of islands in the development of the theory of evolution by natural selection.[42] The logics of the Anthropocene continue this fascination by fomenting interest in islands as embodiments of planetary change and the effects of change. The small island has become a categorical space within GCS scholarship and Anthropocene discourse. Small islands are now exceptional because their vulnerabilities are exacerbated. They are seen as microcosms of planetary change and as exceptional spaces for the study of collapse, adaptation, and experimentation with sustainable solutions. Islands are therefore made into metonyms for the planet itself.[43]

It is common to hear that islands are significant because they are "sites for the emergence of new ways of conceiving the world."[44] This is because islands are distinctive geologic formations, and scientists commonly argue that these formations have influence on natural history and human development as isolated "miniature worlds." This book argues, however, that islands are continually remade to fit this vision. It is much easier to think of cities or buildings as continually reconstructed than to think the same of islands, and yet islands too are designed products.

The significance of islands has been explored by historians of science. For example, in seventeenth-century Europe, islands were imagined as "scientific utopias" where people could control the weather, the sea, plants, and animals through processes of experimentation.[45] European thought about islands inspired the development of the lab concept, wherein controlled practices were envisioned as affecting the very substance of life. Laboratory life was eventually brought indoors, but the efficacy of lab experiments is still based on the idea of island isolation, explicitly subject to controlled manipulation and design. For fields like geology, scientists say that "islands are small, encapsulated units that can be studied and understood in a way that is not possible when we are dealing with the complex interwoven relationships of life on a large continental landmass."[46] This sentiment is echoed by many archaeologists, who consider that "islands are natural history's best shot at something approaching the controlled experiment."[47] Thus, within GCS worlds islands are a "style of looking and inquiry" stemming from a history of laboratory logics.[48]

For many scientists islands are endlessly unique in that they are fragile and one of a kind. But they are also endlessly similar in that they are all potential homes for isolated endemic species and cultures. Observing islands, as a scientist, means observing the universal principles of evolution. Island importance is self-evident in this context—one can *see* the special nature of islands and their importance for understanding Anthropocene realities. Islands are a microcosm of the earth—isolated little worlds in a sea of space—and they are perfect laboratories for the study of planetary change. Within this logic the best way to understand global change is to seek out isolated systems to study ongoing processes. Islands therefore become model systems for simulating the planet as one vulnerable ecosystem, or the "Earth Island."[49]

But some scholars point out that islands are not vulnerable just because they are isolated: in many cases island states are fragile today as a result of past events that forced their *integration* into exploitative global processes. For example, the many islands colonized by Western imperialism are not best understood as discreet spaces. They were forced into world markets that demanded large-scale resource extraction and produced vast social and ecological changes.[50] The point is that island isolation is often more scientific myth than reality, and island populations must not be considered fixed, determined, and controllable; they are in fact part of an expanded world of significant connections.

Climate change is arguably the most well-known small-island issue in the twenty-first century, but climate change reveals the way assumptions about isolation can obscure evidence of integration.[51] For example, the recent association of small islands with climate change is an Anthropocene style of looking. This is especially true at the transnational level of environmental policy, in which the small island has been reified as inherently vulnerable to a host of conditions related to climate change. Island people are said to share this inherent vulnerability by an extension of socioecologic and assumptions about their isolated "islandness." Little attention is paid by the GCS community to the fact that island regions have long been a central part of larger histories of resource use and extraction that degraded island resilience and that helped produce the globalized industrial conditions that led to climate change in the first place.

Because of its historical centrality to the global world order stemming from the transatlantic trade and plantation system, the Caribbean is well known within the humanities as a region of connectivity, not isolation.[52] Yet, within GCS, it is the isolation, vulnerability, and exceptionalism of islands that make them worthy of attention.[53] In this context The Bahamas is seen as an exceptional space that retains functional socioecological systems that other islands have lost.[54] Due to the country's small human population relative to its size, the viable stocks of valuable fisheries species, and intact habitats in the more remote islands, The Bahamas is sometimes referred to as an ecological example of what the wider Caribbean islands "used to be," an example that is now threatened by climate change and other destructive human activities.[55] In this way Anthropocene logics create and compare island categories, and within GCS The Bahamas embodies an exemplary form of Caribbean island socioecology.[56]

Many island anthropologists focus on the possibilities for improving climate change policies that affect islander lives.[57] I focus on the social practices that produce small islands within the Anthropocene, but I also recognize that small islanders are not always treated justly by transnational science and policy.[58] When faced with the island style of looking stemming from global climate policy that either ignores island people or presumes to know what is best for them, islanders may in fact resist and create their own conditions of possibility in the face of climate change.[59] Such imposed islandness and its resistance varies across island regions.

In sum, small islands are significant spaces within the Anthropocene. They have been imagined and designed by GCS research and climate change policy as geologic formations analogous to the earth with particular properties such as isolation and vulnerability.[60] Small islands show how the Anthropocene idea has both material and symbolic consequences and that it is redefining specific locales and geologic features. Small islands have therefore become one location from which anthropologists can now "think the Anthropocene" and study the consequences of the Anthropocene idea. We can no longer think of small islands as simply existing outside of the scientific strategies that construe them as singular spaces. From this angle not only are islands sites that generate thought—*islands themselves* are ways of conceiving the world. But small islands are not the only significant Anthropocene spaces. Other biogeographic formations like the Antarctic, the deep sea, the rainforest, or (de) glaciated mountains have also become laboratory experiments in geoengineering, rewilding, and de-extinction that have new significance in the context of the Anthropocene.[61]

But even if we concede that islands are contingent objects with complex histories, we cannot ignore the significance of contemporary island space making. Islandness, as an Anthropocene style of looking, is an asset for certain ventures. When it comes to scientific research, islands are often referred to as the "canaries in the coal mine" of global change, singled out as indicators of planetary health.[62] Fieldwork in islands is granted a moral significance that translates to a sense of urgency.[63] And for the tourism industry, which is so crucial to the economic survival of so many island states, islands echo the mystique of the sciences when their destinations are designed to sell experiences based on romantic notions of isolation, anachronism, and exploration.

TOURISM AND FIELD SCIENCE REMAKING THE BAHAMAS

> The ties that bind the Caribbean to other places . . . are premised on everyday practices of consumption that occur through economies of movement, touch, and taste in overlapping fields of economic consumption, political consumption, and cultural consumption. . . . In following the tracks of mobility and consumption, however, we must also attend to the things and people that are kept in place in order to enable the mobility of others.
>
> —Mimi Sheller, *Consuming the Caribbean*

One way to observe the Anthropocene in action is to study spectacular events like large-scale geoengineering projects or to examine places that are already showing extreme signs of submersion, extinction, and irreversible degradation. Such things are well worth extended scrutiny. But we must also focus attention on places that, in the process of becoming consumable destinations for "everyday" research and travel, are made to exemplify the Anthropocene. These sites reveal the subtle and pervasive creativity of the Anthropocene idea. The Anthropocene Islands of The Bahamas are one example that has been normalized through the everyday reproduction of consumable place.

Independent since 1973, The Bahamas is a former British colony, a nodal point in the Atlantic slave trade, and a member of the postcolonial British Commonwealth, the Association of Small Islands States, and the Caribbean Community. The mainstay of the postcolonial Bahamian economy has always been international tourism.[64] In 2016 The Bahamas was the ninth most tourism-dependent country in the world per capita, receiving over six million visitors—nearly sixteen times more people than its national population.[65] Therefore, the realities of tourism are essential for any understanding of the function of Anthropocene logics in the archipelago.

Tourism is "king" in The Bahamas and in the majority of the Caribbean region because tourism allows disparate tiny islands to function, providing a legal economy, cash, and needed infrastructure. Tourism is the reason The Bahamas exists today as an independent nation. For the Ministry of Tourism, one of the most prestigious branches of Bahamian government, tourism is more than an opportunity to bring Bahamian citizens into a global capital market. It is *the* flagship industry, and it is normal to hear officials publicly say that their goal is to become one of the top tourist destinations in the world.

The Ministry of Tourism is charged with developing what is publicly referred to as the Bahamian "tourism product." A tourism product is defined within the industry as "a collection of physical and service features together with symbolic associations which are expected to fulfill the wants and needs of the buyer."[66] The Ministry of Tourism extends this definition to encompass the entire country—its islands, cities, and settlements; its ecology, human population, and culture; and its terrestrial, coastal, and marine environment. These must all be managed and sold as features order to deliver the best product experience possible for the visitor.

A significant aspect of tourism mainstreaming involves placing tourism at the center of Bahamian historical and national development narratives. This has gone on since at least the mid-twentieth century, but now Anthropocene narratives are starting to intersect with tourism narratives in official Bahamian discourse. The history of the Bahamian tourism product is increasingly packaged with a history of Bahamian environmental protection: ideas that are often married by the ministry. For example, the 1950s is now seen as both the era of year-round tourism promotion and the decade in which The Bahamas became a self-proclaimed "leader in conservation" in the region. That decade saw the creation of the Exuma Cays Land and Sea Park and the Inagua National Park for Caribbean flamingoes. At the same time, the Bahamas National Trust was created to manage these areas, and together the Exuma Cays and Inagua Parks became what the ministry now describes as "models for conservation" in the Caribbean.[67]

It is therefore possible for Bahamian tourism officials to say that "biological resources are being protected because their touristic value has been discovered."[68] The "tourism product" of The Bahamas is becoming anchored in the nature of islands themselves—islands that have been scientifically identified as vulnerable. And some of the most public rhetoric about anthropogenic island change currently centers around the loss of environmental attractiveness to the tourist industry.

Scientists commonly participate in this focus on tourism and warn about the potential tourism impacts of degraded environments while presenting their work at regional conferences. Many visiting scientists see these arguments as the only way to "reach" Bahamian decision makers they assume are otherwise oblivious to the impacts of anthropogenic change. In its attempt to call attention to fragile island ecologies, the scientific narrative links environmental protection to enhanced tourism. For example, it is assumed that increasing the number of protected areas will attract more visitors, perpetuating the growth of the national tourism product. Here again is the paradox of Anthropocene tourism: anthropogenically vulnerable ecosystems and organisms are protected and promoted as valuable so that their tourism potential can be realized by attracting more tourists to the destination, sustaining the industry while perpetuating its anthropogenic effects. In a sense the Ministry of Tourism and the private tourism sector are using scientific research about threats of island degradation to design means to profit from island fragility.

However, GCS doesn't just participate in tourism mainstreaming through public lectures. GCS comes in a variety of forms, from climate modeling with collected data sets to deep-sea diving in extreme environments, but the meat-and-potatoes research of much academic GCS is based on fieldwork.[69] Fieldwork is the primary technique through which the Anthropocene idea is manifested by scientific researchers.

Within GCS worlds places like Caribbean islands are known for field projects with a tropical small island, marine, or coastal orientation.[70] But such places are also made by research, and what they are made to be determines what knowledge they can produce. For instance, The Bahamas is an "obvious" field site for the study of coral reef–based socio-ecologies, but not for large-scale agricultural production.[71] Attempts to conduct agricultural research on Bahamian islands, in today's Anthropocene funding context, must be framed through reference to island vulnerability, security, and resilience in the face of global change. The international requirements for legitimate field research projects are one of the ways that visiting and resident scholars turn the characteristics of islands in the Anthropocene into opportunities for education, management, and grant-funded socioecological research.

Place is therefore a kind of product that justifies GCS field research, and places are designed through scientific fieldwork. GCS places are valued not only because they can produce geologic or ecological knowledge. Scientists also share the field with many other residents, tourists, migrants, and nonhumans. Field-workers internalize these others into specific places, making them part of their research. In this way island residents become targets for surveys about marine resource use, schoolchildren become the subjects of environmental outreach programs required by funders, and fishers become valuable for having measurable perceptions of local fish species. These people all become a characteristic of a scientifically valuable place.

The place-based products of GCS fieldwork become Anthropocene tourism products when they are taken up through practices of destination design and branding in the international tourism market. This market is more interested in GCS products than ever before. It is now well known that the travel industry—a loose amalgamation of enterprises across scales—contributes enormously to global environmental change through, among other things, emissions tied to the movement of millions of people around the world.[72] This realization about the cost of business as usual has sparked multiple responses, many of which ironically take the form of business opportunities. One example is the packaging of tourist travel with carbon offsetting schemes. Another has been the adoption of carbon neutrality as a destination amenity.[73] And now travel is increasingly directed toward destinations that exhibit vulnerability to anthropogenic global change, change made visible and legible by GCS fieldwork.[74]

The ever-expanding tourism product of The Bahamas now develops places that have been discovered by GCS fieldwork. The industry uses these places to create niche island destinations that can stand out from the standard products of the region. The "sun, sand, and sea" brand belongs to the "Paradise and Plantation" resort model of the twentieth century, a brand that many see as outdated and noncompetitive—a brand for the Isles of June, not the Anthropocene Islands.[75] The advent of global anthropogenic change has become an opportunity for the tourism industry of The Bahamas to reinvent its tourism product, and its island destinations, once more.

The chapters of this book journey across the Anthropocene Islands to describe submerged caves, dangerous fish, remote field research stations, chains of marine reserves, and small-island carbon footprints. Together they reveal The Bahamas as an evolving brand consisting of multiple island destinations. Each chapter presents an example of the tourism industry cocreating viable products along with GCS. "Building Biocomplexity" describes my participation as a social field technician in a technoscientific endeavor I call the "biocomplexity project." Over a six-year period interdisciplinary researchers traveled from the United States to several Bahamian islands to model the country's coral ecology and fishing behavior for the benefit of Bahamian marine managers in the process of creating a new chain of marine reserves. This project exemplifies a shift in research orientation from the ecologic of pristine biodiversity to the socioecologic of global change science. By describing life in small fishing settlements, debates between interdisciplinary researchers, management arguments linking marine protected areas to increased tourism, and the (dys)function of social survey technologies, I show how ideas such as biocomplexity and socioecology are enacted through field research. As a participant in this project, I learned that "the social" has special value for GCS and tourism, but that rural island life does not actually equate with social units like "the individual" or "the community" imagined by such research.

"The Educational Islands" introduces a boarding school for North American high schoolers and a research station for visiting scientists I call the "Island Academy." The academy exemplifies the way research institutions turn the crises of the Anthropocene Islands into travel opportunities. Students come to the academy to experience field education on the island and to immerse themselves in island environmental studies, becoming inculcated as Anthropocene subjects. The island itself, including its people, is designed to be the "teacher" for these students. But there are many connections that are hidden from students to enhance the "islandness" of their experience. By making the island into a grand educational fieldtrip, the academy plays down the role the students and their families play as consumers of postcolonial place in the tourism industry, and the complex connections the school has with the island as a multinational enterprise are not incorporated into the curriculum. I found that the narrative of island isolation obscures both the academy's and the island's connections to other places, people, and historical events even as the island is presented to students as a microcosm of global change issues.

In "Sea of Green" I show how tourism officials attempt to make climate change a marketing asset for an island nation within the Anthropocene logics of vulnerability. This chapter describes a climate change and tourism workshop in Nassau and a presentation proposing that the Caribbean become the first carbon-neutral tourist region in the world. The presenter advocated marketing carbon neutrality as a desirable condition to maintain regional distinction. Despite the fact that travel emissions are the greatest way that the island region contributes to the buildup of greenhouse gases, tourists might be enticed to buy a ticket for a trip that promotes a green consciousness. The irony is that continued mass tourism trumps every conservation-oriented plan in The Bahamas, and yet conservation can be repackaged by the tourism industry to continue to bring visitors. This is an example of the paradoxical alignment between the global climate governance regime and international travel markets.

"Aquatic Invaders in the Anthropocene" shifts to the production of fisheries within the management logics of the Anthropocene. I introduce a young Bahamian scientist studying the habits of the lionfish, a popular aquarium commodity and invasive species in the Bahamian coral reef system. She was measuring the level of vulnerability of Bahamian waters to lionfish invasion, a project designed to operate within the logics of GCS. This chapter also introduces the national lionfish management plan to entice Bahamian fishers to catch, clean, and sell the invasive fish to residents and tourists as a sustainable commodity. Making a commercial market for species eradication makes sense to marine managers because fishers are represented in the scientific community as chronic overfishers. I show how both fishers and lionfish have been paradoxically positioned as threats and saviors, in part because they are both closest to the coral reefs perceived to be in danger from pervasive anthropogenesis.

In "Down the Blue Hole" I return to the underground, undersea blue holes to demonstrate how GCS produces new forms of national history, heritage, and belonging through the re-creation of newly valuable landscape features. I relate my trips to the islands of Andros and Abaco, where I assisted Bahamian college students and U.S. scientists in a research project to investigate the caves. Foreign biologists, paleontologists, and archaeologists had been searching in the holes for preserved plant, animal, and human remains to learn more about planetary change and to link scientific data with the social uses of blue holes. This chapter describes the way rural Bahamian residents separate themselves from the current study of the holes, assuming that scientific research makes the holes useful objects for island tourism and therefore sites for foreign visitors. I explain that the more scientists found in the watery holes, the less people considered them part of everyday life, even though they represented a source of income. I show how blue hole exploration has become a selectively synergistic collaboration between tour operators, local island government, national heritage institutions, and scientific disciplines searching for clues to the Anthropocene.

I conclude with "Anthropocene Anthropology," a call for more scholarship about all the places undergoing symbolic and material re-creation within this era of anthropogenesis. Arriving at destination Anthropocene means nothing more or less than reimaging the parameters of the living world, and I promote anthropology as an essential tool for exploring the power-laden reimagination of the planet. While this book is grounded in the specific circumstances of the small islands of the twenty-first-century Bahamas as they are remade by the Anthropocene idea, GCS, and the tourism industry, different intersections of science, capital, and postcolonial politics are involved in similar transformations of other places around the world in the name of the Anthropocene.[76] As the science of global change advances and expands, changing not only how we understand the world but also how we live in it and how we design possible futures, it is my hope that this book will inspire future researchers in other destinations to reveal the Anthropocene idea in action.

I care deeply for The Bahamas on two levels. I care about the country personally, because I am richer for the experiences I have had and the lifelong friendships I have made while living and working there. But I also love the islands of The Bahamas anthropologically. Anthropologists have a fascination with the Caribbean because the island nations of the region continue to reveal how the contemporary world is shaped by overt and subtle forms of power.[77] It is this love that inspires me to tell the story of The Bahamas as an anthropologist of the Anthropocene. While not everyone will experience the same personal joy in the archipelago as I do, the anthropological appreciation of the place is ultimately more captivating and more instructive for all of us who wonder what the Anthropocene idea means for human and nonhuman lives everywhere. Long-term, qualitative, multisited research among the emergent projects that are busily remaking the places we love is crucial for publics even outside of anthropology. This research deepens our collective understanding of current events and retells familiar stories from new angles, allowing for other possible outcomes. This kind of engaged research should matter for all of us who share the planet.

1

Building Biocomplexity

BEGINNINGS

I first traveled to The Bahamas in the summer of 2002 to work on an interdisciplinary research project funded by the U.S. National Science Foundation (NSF).[1] I was a twenty-one-year-old undergraduate, and my assignment was to work with a team of U.S. researchers to administer pilot surveys to people we had never met in a place most of us had never been. Our fieldwork took place in Cherokee Sound on the southeast coast of the island of Abaco, a small settlement of fewer than two hundred households. In Cherokee we administered surveys, studied seasonal fishing practices, and documented the patterns of daily life we were fortunate enough to participate in for a few short weeks. At that time it seemed the kind of one-off international visit—a kind of "study abroad" program—that often accompanies U.S. higher education. Yet many years later, after many trips to The Bahamas, I vividly remember that summer. It was a kind of beginning.

Cherokee was selected by the study designers for a variety of reasons: its history of settlement "interaction" with the marine environment over generations of fishing and boat building; its proximity to a range of what the project termed "ecological zones," including the deep-ocean mangrove stands, coral reef systems, blue hole caves, shallow-water flats, and pine forests; the community's shared descent from white British Loyalists; and the fact that few tourists visited and few people moved away from the settlement. I was told this meant that most interviewees

FIGURE 2. Satellite image of the Bahama Islands. NASA image created by Jeff Schmaltz, MODIS Rapid Response Team, Goddard Space Flight Center (2009).

were representative of the community and had not been greatly influenced by life outside. Cherokee was a model island fishing community. I would soon find, however, that these parameters explained very little about life there.

Two brothers, Sam and Steve, stood out vividly in the settlement. They were often found on the small wooden dock adjacent to Sam's house.[2] I was shy and unaccustomed to their sharp-eyed scrutiny, wry laughter, and sun-leathered skin. Their dock was worn and ramshackle, and when I first met them I had yet to recognize that the stained structure built of scavenged boards and plywood was a fish-cleaning table, scoured by seasons of use. Now tables like these are one of the first things I look for as a marker of an active fishing settlement.

My notes detail walking along the nearby shore, following children visiting from Nassau. Because they summered in Cherokee with grandparents, they were familiar with littoral and mangrove life. The loved to chase "bonga" in the shallows, a small and darting fish that blew itself up into a great round ball when startled. They also showed me how to excite the snapper schooling at the end of the settlement's long dock. Black bars appeared on the heads of these fish when feeding, and the children often threw them bread to spark this transformation.

Another fisher named Charlie spent his summer laying fish pots, collecting conch, and spearing grouper and hogfish on the reefs. He was loud and easily amused, often walking barefoot and shirtless around the settlement, perfectly at home. He sold some of this catch locally, ate some of it, and gave a lot of it away to neighbors. It was from his boat that I caught my first small reef fish while handlining over the side, and on his boat that I first learned the local names for the species of snapper, grouper, and reef fish that aggregate on the coral heads around the island. At that time Charlie was in love with a visitor from the United States, a woman who seasonally rented one of the few guest houses in Cherokee with her two children. Her children adored him. He would take them fishing, cooking their catch for dinner.

Though he fished every day, I soon realized that Charlie was on vacation. Most of his summer fishing was pure pleasure. For Charlie these summer days of small boats, fish scales, and conch shells were a way of life but not a living. Commercial crawfishing on a cooperatively owned smack boat between August and March was his primary vocation. A number of men in Cherokee were also professional crawfishers, leaving home in the season to join their crews on an island to the east, although none were as garrulous and generous with their summer time as Charlie.

FIGURE 3. Fish in a bucket, Abaco. Photo by author (2014).

My experience as a research assistant influenced the next phase of my adult life, aligning my future with The Bahamas. My memories are entangled with my field notes from these early years, and I present sketches from those notes here to show how Bahamian island life diverged in form and content from the methodological ordering of the science project I worked for. I had my first taste of global change science (GCS) research on this project while simultaneously awakening to an engagement with sea-based lives. And it was this project that made me skeptical about the capacity of common social-assessment tools to capture the dynamism of the world.

The research project that began in Cherokee Sound was one piece of a larger endeavor designed to study something called biocomplexity. As a scientific term, *biocomplexity* refers to the complexity inherent in the structure, functioning, and relationality of all living things across scales from cells to microbes to organisms to populations and finally to entire socioecological systems.[3] As a concept within the natural sciences, biocomplexity is seen as a contemporary update to the concept of biodiversity, which more narrowly describes "the variety of plant and animal life in the world or in a particular habitat, a high level of which is usually considered to be important and desirable" for conservation.[4] Biodiversity, as a concept that implies there is tangible evolutionary and even commercial value held in life forms, has been used to extend the reach of international science and capital in the world.[5] And now biocomplexity projects are one of the means by which GCS field research creates research objects amenable to environmental management and commercial enterprise.

This chapter describes the biocomplexity project as a field-based experiment placing scientists and field technicians in situ to approach Bahamian socioecology at the local level. I discuss one of the ways that biocomplexity projects can create the human social worlds they study in the name of social inclusion in ecological research. These social worlds become simplified caricatures of actual social life. Such reductive social caricatures align all too easily with Anthropocene spatial products such as chains of marine protected areas (MPAs). These products have in turn been shaped by prevailing assumptions about tourism and resource management. The end result is that the local people studied by such projects run the risk of losing the richly meaningful relationships they have with the ecology that surrounds them.

SURVEYING

Back in 2002 my activities as a social science field technician on a biocomplexity project consisted of performing various "practices of place."[6] These acts included maintaining field notes about settlement life in Cherokee, participating in community activities such as fishing, and conducting structured interviews with human subjects using a preset survey questionnaire. The survey centered around the ways local people individually interacted with the environment, the costs and benefits of doing so, how they thought the local environment changed over time, and their perceptions of marine regulations. My team administered this complicated pilot survey to willing, if slightly bored, Cherokee fishers and their families. After a day of surveying, we compared our experiences around the kitchen table of our rented guest house.

The field team for this pilot study consisted of three young U.S. women in our twenties. We were led by a seasoned field anthropologist and academic who knew Abaco and who had helped design the project with an interdisciplinary team of researchers from several research institutions in the United States and Europe. It was on this trip that I first experienced the ennui that accompanies repetitive field research in an unfamiliar place. The pilot survey instrument—a list of typed multiple-choice questions and scales we carried everywhere we went—became a tool that performed a number of tasks for its users. The survey shielded us from feeling like tourists, but it also put up a barrier between us and the people in the community. It signified that we were not like them because we were there to study them. The survey isolated us from the community, in some ways more than our initial strangeness as foreign visitors.

This biocomplexity survey was cutting-edge for its time. It was an attempt to manifest Anthropocene socioecologics about a given place in quantifiable form.[7] It did this by modifying the standard template of a household socioeconomic survey to include questions about specific forms of "interaction with the environment."[8] The social survey was deemed progressive by the project designers because they felt that pre-existing ecological assessments conducted by natural scientists had not adequately incorporated social data, and most social surveys conducted by social scientists did not adequately incorporate the ecological perceptions of respondents. The methodological novelty of this survey, at that time, was the combination of these concerns in one interview template and the integration of the social survey data with ecological data collected in the country as part of the same project.

The survey tool introduced a new form of ecobiopolitics to The Bahamas.[9] Within the survey, interaction with the environment was defined primarily in terms of the targeting of specific species while fishing ("What species do you primarily fish for? When do you fish for that species? Do your target species change over the course of the year? What are your target species in each season?") and secondarily in terms of engaging in specific outdoor activities ("Do you ever engage in the following activities: swimming in the ocean, walking along the shore, collecting in the mangroves? How many times a week do you engage in those activities? Rank these activities in terms of frequency and enjoyment").[10] While there was room for writing in "other" entries on the survey, the majority of response options were predetermined based on the survey creator's prior research with fishing communities elsewhere. The survey also asked about the legality of fishing practices in terms of knowing, understanding, and obeying fishing regulations, eliciting responses about the awareness of MPAs as a key form of marine management ("Have you ever heard of a marine protected area? Do you know if there are any nearby?").

In terms of practical use, the survey format was highly social. In Cherokee the project's need to survey as many households as possible drove us to approach people we might never have encountered otherwise. In pairs, but most often on our own, we would cautiously approach the front door of a brightly painted house. As welcoming as the colorful paint was, the gates often gave us pause, as did the growling, skinny "potcakes" (a Bahamian breed of street dogs) accompanying some yards. Once we were safe inside, the extent of the forty-question survey meant our encounters were lengthy, causing some respondents to grow increasingly comfortable with our unfamiliar presence in their home. On some visits we sat on floral couches with elderly women, answering questions of our own they posed about our homes in the United States, our parents, our boyfriends, and what it was about The Bahamas we most liked. But while survey questions about illegal fishing practices made some respondents uncomfortable, when I say that the survey isolated the field technicians from the community, I mean that the questions themselves positioned us as interlopers who could learn only what we already wanted to know.

In the summer of 2005 I traveled to The Bahamas to administer another round of surveys for the same project in a new settlement. The survey format had been finalized, and the new location was one of several ultimately surveyed by the project. This time I was sent to the settlement of Tarpum Bay, a fishing community on the island of Eleuthera. This settlement was chosen for its proximity to a prospective MPA that was to become part of a Marine Reserve Network in the country. While Harbour Island at the northern end of Eleuthera is a well-known tourist destination, Tarpum Bay is relatively far from the tourist path. Again the team administered surveys, practicing as random a sampling method as possible in a small community of 230 households. We entered survey data on-site into a computerized database, discussed sampling strategies, and shared information about our daily excursions from our guest house into the settlement.

Yet again the survey created encounters that might not have otherwise taken place. We learned that Traveler's Rest—a wooden shade structure next to the town dock with benches positioned for the breeze—was a prime spot for surveying. We could ask our questions as fishers cleaned their day's catch, brushing stray scales off the survey and out of our hair as we went through the questions. And we had an excuse to hang around the pizza parlor, hoping to survey anyone waiting while their order was made. Our novelty offered more entertainment than the parlor television. Again the survey instrument was both the excuse that validated our presence and the embodiment of our separation from the people of Tarpum Bay.

These trips to the Family Islands of The Bahamas were a taste of GCS research in the Anthropocene, though I wouldn't have thought of them in those terms at the time. I had volunteered to be a "technician of general ideas," who could test a language for reimagining the relationships between people and their environments.[11] These experiences were awkward in the sense that we struggled along with our subjects to understand what the survey questions meant (What is an MPA? What kind of habitat is "hard bottom"?); what answers were acceptable (how frequent is "oftentimes"?); and which stories were relevant to the project and which were not ("The Americans keep coming here with their boats and poaching our fish"). Our desire to please both the project designers and the research subjects meant long periods of wrestling with the software and long conversations about the meaning of responses. We learned that the answers to our questions could not be easily transformed into preset categories of the survey. Making our conversations with islanders "empirical" meant transforming their wide-ranging stories, asides, and cautious statements into "data" that could be collated and compared. We walked the knife edge between artifice and science that lies at the heart of all fieldwork.[12]

During these field seasons I began to ruminate on the survey tool and the relationship between surveyor and surveyed. On one hand, field technicians have little time to develop rapport between themselves and the people they survey because there is pressure to collect as many surveys as possible in a short time span. Researchers can feel conspicuous, foreign, and presumptuous for approaching local people and asking them to participate in the project and sign the consent forms mandated by U.S. human subjects–research protocols. On the other hand, surveys bring researchers into public meeting places and private homes, they announce researcher presence in settlements, and they open doors into arenas of public controversy that might not otherwise come up in conversations between citizens and visitors. As representatives of scientific projects that target communities, visiting field research technicians are a perpetual part of the tension between curiosity and resentment these projects inspire in islanders.

These early biocomplexity project experiences in The Bahamas would come to shape both my interest in contemporary anthropology and my ideas for this book. Through this field project and its survey technology, I began to recognize the subtle ways in which The Bahamas was scientifically manipulated for the Anthropocene. I also began to see how these projects could lay the groundwork for the creation of Anthropocene tourism products.

BIOCOMPLEXITY IN THEORY AND PRACTICE

How did biocomplexity become a concept that launched years' worth of U.S. social survey collection in small Bahamian settlements, transforming the archipelago as a destination for interdisciplinary socioecological research?[13] To answer that question we must begin far from the Bahama Islands, in Baltimore, Maryland, in 1998, at a plenary session of the forty-ninth annual meeting of the American Institute of Biological Sciences.[14] Rita Colwell, the director-designate of the NSF, addressed an audience of researchers, describing biocomplexity as a major new research agenda. Colwell explained biocomplexity as the twenty-first-century response to the increasing vulnerability of the planet in the face of anthropogenic degradation. She defined the term as the chemical, biological, and social interactions of the earth's systems that must be studied to solve challenges for future planetary sustainability.

Anthropogenic environmental crisis, in this frame, evokes scientific responsibility but also an opportunity for the sciences to extend their horizons. Biocomplexity was touted as the "key to social understanding" that had been missing from prior NSF-funded environmental research. Colwell contrasted biocomplexity to biodiversity, situating biodiversity as the research paradigm of the 1990s. She recognized the importance of sustaining biodiversity by protecting plant and animal species, but moving beyond species to incorporate human and biogeochemical processes into research was a deeper concept for Colwell, one she wanted to promote as widely as possible.[15]

In 1999 the NSF began funding research projects under the category of Biocomplexity Research, and in 2001, at another American Institute of Biological Sciences annual meeting, biocomplexity was refined further for a curious audience. Panelists proposed a tentative definition for the term, with the presumption that this definition would be modified in the future. Biocomplexity was "properties emerging from the interplay of behavioral, biological, chemical, physical, and social interactions that affect, sustain, or are modified by living organisms, including humans."[16]

At the turn of the century, biocomplexity defined a new mission for bioscience. In its promotional materials, the NSF explained that breadth and complexity were exactly what had been missing from the life sciences. These fields had become ever more reductive, emphasizing specialization as a result of tight monetary constraints. Complexity, in the prior research environment, had been an obstacle to overcome, and vast efforts had been expended to reduce complexity to studies undertaken by lone scientists or small groups of researchers. The implication was that all this reductionism informed the majority of scientific knowledge to date, but these projects had not yielded robust information for "real-world" problems defined by their complexity and uncertainty—what would soon be described by some earth scientists as Anthropocene problems.

Biodiversity, as an earlier research paradigm, was configured by the NSF as having devolved into the cataloging of biological data on specific organisms, providing few hints of the relevance of this data within larger systems of global interaction. In other words, biodiversity, as a concept, provided no substantive sense of systemic relationships or emergent properties. But planetary systems were said to be in crisis in the Anthropocene, threatened by intensified population growth and pollution, becoming increasingly vulnerable. Biodiversity, as the planet's valuable collection of genetic information housed in species, was theoretically reframed as *sustained* by biocomplexity, and ecological research concerns at the NSF were switched from the enumeration of biodiversity to a broader focus on emergent processes, including human activities as an essential component.

As a GCS "technobiological imaginary," biocomplexity is more than a theory. The concept focuses explicitly on the method and practice of GCS research itself.[17] To understand the complex interrelations of living systems, the logic of biocomplexity requires that scientists reach across disciplines to study biological, chemical, and social interactions. Interdisciplinary collaboration is assumed to engender more robust forms of knowledge production, and thus interdisciplinarity is at the root of biocomplexity research practice. In sum, living interactions in the real world can be understood only by scientists from many disciplines who work across scale from the macro to the micro. Mathematics is considered the most fundamental language for understanding biocomplex systems because it is considered to be the only scientifically universal language.

My interest in biocomplexity as a form of Anthropocene socioecologics briefly took me away from The Bahamas to College Park, Maryland, for an interview with Dr. Rita Colwell herself on a chilly day in 2008. Dr. Colwell received me graciously, making sure her assistant offered me tea and cake. After ten years she was still quite proud of this paradigm shift at the NSF. She explained that biocomplexity was a conceptual and ethical triumph because the research was "much more than an ecology program." It was a program that could contribute to real political problems centered around risky planetary futures. And the idea came with a doubling of the NSF budget, a major feat at the time. Colwell attempts to communicate that "we are part of an extraordinary adventure—and there are things we can do to maintain the gift of life we have been given." She was pleased that biocomplexity was still in existence as a major NSF program, promoting interdisciplinary innovation under the new name, Coupled Natural and Human Systems Research.

After sifting through the literature and speaking with Dr. Colwell, I now see biocomplexity as a logical step in Anthropocene reason beyond diversity. In 2008 Dr. Colwell hoped that biocomplexity research might culminate in an ideal earth model wherein all significant biological, geochemical, and social interactions could ultimately be predicted. Biocomplexity is therefore an experiment for GCS decision making in the Anthropocene and an attempt to combat the proliferating uncertainties stemming from the extreme anthropogenic drivers in the earth system. But the concept exacerbates a number of tensions when put into practice.

The Bahamas project I participated in was one research endeavor that successfully met the NSF's funding criteria. It became a multiyear project that mediated between two transnational arenas. The first was the NSF's Biocomplexity Research program, as just described, with its concern for interdisciplinarity and the production of socially robust systemic models. The second was the ongoing and highly political marine conservation scene in The Bahamas.

In 2000 the Bahamian government announced its intention to create a Marine Reserve Network, which would initially include five MPAs. These areas were to be designated as "no-take" reserves, areas that prohibit the extraction of marine resources, and they were a response to the overconsumption of key marine species. The announcement came after two years of planning meetings between The Bahamas' Department of Marine Resources and environmental nongovernmental organizations, including the Bahamas Reef Environment Educational Foundation and the Nature Conservancy of The Bahamas, who feared that sustained overfishing would lead to the destruction of the Bahamian coral reef system, overall biodiversity loss, and declining fisheries productivity. These groups predict that without some kind if intervention, The Bahamas will go the way of the majority of the Caribbean and lose species diversity and valuable commercial fish stocks.[18]

The biocomplexity project was a loose entity made up of U.S. and European researchers from fields including anthropology, biology, oceanography, physics, economics, and mathematics. The project's research was intended to inform the planning of these marine reserves. Within the life sciences and environmental nongovernmental organizations it is commonly assumed that MPAs are an ideal milieu for the integration of social science and biology.[19] The project successfully proposed to model MPA feasibility and the systemic effects of MPA creation to produce recommendations for the Bahamian government, as well as simulations of coral reef socioecological functioning.

There were three main research components within the project: the social team, the habitat team, and the genetic connectivity team. The stated goal was for each team of researchers to collect data separately and then to integrate data across components to produce holistic human and environmental models of the Bahamian reef system. The social team went about "assessing patterns of resource use and attitudes about resource conservation among stakeholders" and designing the social survey technology to compile comparable data sets from settlements situated near proposed reserve areas.[20]

The survey my team administered was concerned with statistically elucidating the connection between economic conditions in a settlement and the intensity of marine resource extraction conducted by individuals within that settlement, as well as what respondents thought about the appropriateness of such extraction—all part of the project's effort to model human and environmental interaction. Anthropology and economics, disciplines representing the behavioral sciences, were enlisted to make sure the knowledge produced for the modeling project reflected the sociality of The Bahamas. Anthropology and economics were seen as disciplines that would legitimate social claims made by the project.

Some of the findings of the social team were eventually published in an article that stressed the necessity of social assessment for resource management and governance. The authors quantified variables and environmental perceptions held by two kinds of entities identified by the project: individuals and communities. Variables, for these social scientists, are traits that can be pinned to particular individual or community entities and then compared across a number of individuals or within communities. Using the results from the field survey template, they assessed specific demographic variables of respondents, including age; number of children; level of education; marriage status; gender;

occupation as either tourism, fishing or "other"; and household income. They also asked if the mother was from the specific settlement, if past generations of their family had been fishers or farmers, if they had heard of marine reserves or been to a reserve meeting, and how frequently they went to the sea to use marine resources. These variables were calculated for five island communities across hundreds of survey participants.

Social perceptions pertained to participant responses to environmental management ideas such as the state of local marine conditions, the level of threat to the marine environment, and the state of the enforcement of fishing regulations. These "perceptions" were paired with variables such as household income, reliance on fishing income, reliance on tourism income, and whether the participant thought there should be a local marine reserve in the area. The demographic variables were described in the article as material aspects of life while the perceptions were described as individual and community "perspectives."

When statistically linked, certain material aspects of life for an individual or community can be shown to influence certain individual or community perspectives on the marine environment in a given place. The assumption is that human and environmental interactions are shaped by individual perceptions based on the material realities of life. This data, when collated for specific communities, can, in theory, predict what perceptions local people are likely to have about MPAs given the specific material conditions of that community, thereby potentially predicting the success or failure an MPA might have as a management strategy in that place. This kind of reasoning has become increasingly common within the social assessment branches of GCS projects.[21]

ASSESSING SOCIAL ASSESSMENT

Biocomplexity exemplifies the way "the social" is produced within GCS projects in the Anthropocene. Within international environmental research and governance regimes, social assessment is popular because it can encompass almost anything related to the study of human life.[22] The reasoning is that "social" phenomena are not necessarily as specific as "cultural" phenomena. Social research and social data are considered to be more generalizable for cross-scale management strategies, translocational comparison, and social expertise stemming from any social scientific discipline or method. The "social human" is assumed to be universally assessable.

The social survey assumed in this case that notions like individuality and community represented Bahamian island sociality, because biocomplexity research promotes the notion of the social, requiring social data to become a substantial aspect of field research methods and questions. The assumption is that sociality exists in a given research site that can be assessed, standardized, and formalized within an interdisciplinary research project. This requirement means that the shape sociality takes within GCS biocomplexity projects is often predetermined—it is *recognizably assessable* in the first place.[23]

One of the defining features of the survey was the categorization of each person interviewed as an individual with a livelihood. This meant that individuals were assessed by their occupation, categorized into either tourism or fishing. The occupational history of their parents and grandparents was also documented. This is where the project's assumptions about sociality become explanations of local islander behavior, wherein occupation is tied to particular extractive activities, appearing later in the survey, involving notions of self-interest centered around livelihood.[24] These behavioral explanations are based on assumptions that individuals extract value from the material environment through their jobs and that employment can distinguish one person from the next and one settlement from the next in terms of perspective and perception. People with similar jobs were then lumped into "stakeholder groups," and stakeholder groups came to stand in for more complicated relationships and histories.

The key is that the social survey is a tool used here to create categorical distinctions and stakeholder groups suitable for targeted management. In the case of this biocomplexity project, survey categories based on individual occupation created the stakeholder groups "commercial fisher" or "tourism employee." Survey clusters based on settlement location became "communities" that had their own calculable traits depending on the stakeholder groups within them and the behaviors linked to those groups. Such categories are not "wrong." They do approximate a version of events even as they produce those events and the actors in them. Yet I remember Cherokee Sound and Tarpum Bay differently, in spite of the fact that I was directly involved in collecting this social data.

Charlie helps bring the Cherokee of 2002 back to life in ways that defy the easy categorization of "commercial fisher" or "fishing community." My notes are a jumble of information about his ideas that didn't fit into the survey. For example, Charlie liked to talk with the researchers about fishing regulations. He approved of the proposed closed season for spawning Nassau grouper that the nation's marine managers, based primarily in Nassau, were trying to impose because he recognized that the fish were in decline.[25] A catch of twenty-seven grouper is an unusually large number, and the rumor of a local catch of this size lingered in his thoughts. Yet he summarized his personal ecology as a hunter by saying, "Fish are practically like vegetables," while puttering on his twenty-five-foot boat that housed two plastic chairs, one engine, a broken radio, and a barely functioning CD player. He informed me that he wouldn't spear undersized reef fish unless using them for bait, and he said in response to my vegetarianism, "I suppose you think Bahamians are cruel (to fish), but its just a way of life."

My notes also recorded Charlie's more remarkable activities. To collect queen conch, large and sweet-tasting mollusks, Charlie liked to let out a rope behind his boat, hold on to it with one arm, and drift well behind the churning motor, while wearing a snorkel mask.[26] He would look down on the sea floor as he was slowly towed over sea grass beds and the hard bottom and yell out to his driver if he passed over promising conching grounds. On one trip he collected twenty-three mature conch. He taught me that they can be kept alive in captivity for days at a time if placed in submerged conch pens at the edge of the sea.

Charlie helped our team have experiences I will never forget. On one of the last days before we left for the United States, the residents of Cherokee Sound threw their annual summer picnic. Charlie took our team of young Americans on a quick boat ride up the coast to the party site in a horseshoe shaped bay. This was one of his favorite places, and we immediately understood why. The beach was undeveloped, the dunes covered in waving sea grass and the tall stalks of century plants. Cherokee residents often used the site for social events, and the community played beach volleyball and ate peas and rice, fried snapper, and conch fritters at sun-bleached picnic tables. Charlie joked with everyone, proudly watching over his girlfriend's children as they swam. Someone brought a stereo, and music played throughout. We all stayed until the sun went down and the stars came out. The bay has since been developed by a luxury multinational hotel brand, a private golf course has been installed, and Cherokee residents no longer use the beach for summer parties.

My field notes from Tarpum Bay in 2005 are a similar collection of flotsam encounters that could not be included in the survey database, describing blood on the dock during fish-cleaning sessions and laughing gulls swooping overhead in their dark summer plumage. Fishers showed me how to hold fish by the eyes to move them around the cleaning

FIGURE 4. A shark swims under a dock, Eleuthera. Photo by author (2014).

board when their shimmering skin was too slick to grab by hand. A particularly spry old man in the settlement, named "T," volunteered to take me out in a little wooden-flats boat he made himself to handline for permit and mutton fish in the shoals. "You know a storm is coming if you look over the side, and the fish are showing you their white bellies," he said. He filled my head with stories of a giant moray eel that lived in a hole offshore from the settlement. When I recounted these tales to younger fishers, they laughed uproariously and said that T was pretending he was back in the island's tourist heyday of the 1970s, trying to impress foreign girls with his local knowledge. That heyday is long past.

I have snatches of notes from those trips with the biocomplexity project that detail children playing on Sundays off the end of the settlement docks. These children paid no mind to small sharks while swimming but baited hooks for sharks at night as a form of sport. Young men shot white crown pigeons, hunted wild hogs with dogs in the pine forests, and speared fish in the mangroves, but they also occasionally built guest houses, serviced yacht marinas, and did limited contract work at the few nearby resorts: an endless array of activities that cannot be reduced to "local interaction with the island environment."

The biocomplexity project concluded data collection in 2006. In the years since I have come to believe that it is not just occupation that best explains social processes in The Bahamas. Rather, it is the scientific *assembly* of interests, values, and interested persons that is itself worthy of study.[27] This awareness allows my focus to come off collecting and comparing the behaviors of local people within a given occupational category. My focus can shift to the assumptions held by the research project itself and the way those assumptions about sociality actually limit the scope of social assessment. The project's ideas about individuality, community, value, and perception then become anthropological research objects themselves.

Such a shifted focus helps demonstrate that there can be no assessment of research subjects without simultaneous processes of subject formation.[28] I now see the individuals and communities (the "local stakeholders"), defined by occupation and location, as *produced* by the project, not explained by it. Such a shift reframes the social survey as a tool for inclusion and exclusion that dictates the ways certain people can be recognized and assessed within biocomplexity research.

The surveys employed by the project activate figures of the local and rural and of the Bahamian Family Islands through occupationally evaluative variables that come to stand for a sort of personhood. "Fisherman" becomes an occupational category that signifies extractive activities for self-interested gain and claims to subsistence tradition that are different from activities connected to the category of "tourism employee." Tourism and fishing become construed as existing in an inverted relationship. For example, fishers were statistically shown to be less likely to support marine reserve creation and tourism employees more likely, based on what are described as different forms of interaction with the marine environment linked to diametrically opposed perceptions of that environment.

The social survey has become the primary tool for social diagnosis in Anthropocene GCS, wherein research subjects can be only objects of study, prevented from acting as authors, and their participation becomes drastically proscribed.[29] Further, field technicians are also produced as objects of the survey, standing in as representatives of the double legitimacy of social scientific research and the supposed transparency of the survey method itself.

Away from "the field" the strategic production of island sociality became evident to me when I attended a general meeting of the overall biocomplexity project in early 2007 at a university in California. The designers of the research across all three teams (social, habitat, and genetic connectivity) attended the meeting. At times there were nearly twenty people present for team presentations, and over three days these

GSC scientists discussed the possibilities for data integration, holistic model making, and the overall results of the study.

In a fluorescent conference room over coffee served from a warmer, the members of the connectivity team explained that marine reserves preserve the genetic diversity of marine organisms when designed around key populations of marine species. They concluded that networks of reserves were better for the preservation of species with large dispersal areas. But they were uncertain if these conclusions, based on computer-simulated models, could hold up to real fishing pressures and dynamic fishing behaviors. The habitat team was equally certain that habitat areas were decent surrogates for fish diversity. Their models showed that the greater the diversity of sea grass beds, mangroves, reef patches, and elusive hard bottoms enclosed in marine reserves, the more likely these reserves would be to preserve the most diversity of marine life. But they were equally uncertain about the effects of fishing pressure and practices on these models. Certainties about biodiversity conservation vanished when faced with the prospect of confusing social realities. The natural scientists were reluctant to oversell the conservation benefits of marine reserves in a dynamic fishery that fluctuated by season, species targeted, rate of fishing activity, and method of fishing.

The designers of the project had initially proposed to holistically model this complex coral reef–based system, using the language of statistics and mathematics, to create "reserve selection algorithms" enabling policy makers to make well-informed social and ecological MPA-siting decisions. The idea was that an algorithm based on real data could meet conservation goals with minimal costs and identify the ideal locations for reserves. But after five years of data collection, the teams were at an impasse between creating a tool that prioritized modelability or ceding to the slower political process of creating reserves that could weather fluctuating fisher practices, local rights, and ecological change. One scientist said ruefully, "We are like consultants trying to create a package solution for The Bahamas." Another conceded that the creation of one functional model, as had been originally proposed, was not going to be feasible. There was also confusion about who all this research benefited. One researcher opined that the research should be for government decision makers. But another researcher countered that this research should be aimed at settlements because it was the fishers, after all, who had to agree with the conclusions of the study to support any recommendations that might stem from them.

The social team did publish their quantitative survey data, but they acknowledged that this data could not be easily integrated with data collected by other teams. The other teams needed the social data to come in the form of "driving variables" that could explain or "force" the calculated relationships within the ecological models, and this data was not forthcoming. The social data had not been collected in the same locations as the ecological data, and it had been collected at the household level, which was not comparable to ecological data collected at larger scales. It could not readily align with the connectivity team's information about genetic movements of key species or the habitat team's information about the spatial diversity of marine environments. The fact that the ecological teams did not seem interested in listening to the social team's concerns led one member of the social team to complain in a breakout meeting that "no one in these kinds of projects cares about social data. This project is all about modeling and nothing else." Another member agreed and cautioned that small fisheries are often drastically affected by marine reserves in their fishing areas and that the loss of fishing rights can kill communities that have historically depended on local waters for livelihoods and subsistence. But, despite these internal tensions, the project's public face still promoted the scientifically informed siting of MPAs.

This general meeting helped me see that interdisciplinary biocomplexity work is messy. Problems that seem so settled in research proposals are often far from neatly defined in practice, and project success can be based on far more than significant research results. For example, each team's disciplinary work on connectivity, habitat, and social perceptions advanced scientific careers and produced publications. It doesn't matter in the end that much of the work was incommensurable. The work was funded because biocomplexity research was of the moment and because manifesting "the social" has value of its own within that framework.

The biocomplexity project exemplifies the way social science is performed to legitimize it within GCS practices in the Anthropocene. By performed, I mean that the social aspect of socioecological research projects must be demonstrated in ways that are easily identifiable as social.[30] The project produced a form of assessment that does this work of social demonstration to turn local people into familiar categories for use in an environmental management field that perpetually promotes MPAs as the most valid form of marine spatial governance.

As a participant in this project, I am now more cautious about the techniques that validate environmental management. It is one thing to show that there are a diversity of perspectives held by rural Bahamians about the marine environment. But it is quite another to recognize only the kind of diversity that can be manifested through preexisting categories. This is a proscribed form of diversity. The result is not a deeper understanding of what it might mean for islanders to hunt, fish, eat, flirt, raise children, adapt to fluctuating tourism markets, and reinforce community in an ever-changing land and sea world. Instead, "social assessment" has become a tool to discipline anthropogenesis by targeting specific communities for management interventions. In this case the management intervention was the creation of MPAs suitable for the expansion of marine tourism instead of fishing.

BUILDING BIOCOMPLEXITY IN A CARIBBEAN ARCHIPELAGO

Like many ventures in Caribbean history, the biocomplexity project attempted to remake selected island communities into experiments with ecobiopolitical management.[31] In this case the project attempted to remake the region into a model system, proscribing forms of sociality within GCS research as well as the regulatory and livelihood options available to those participating in the program as newly formed "stakeholders." The biocomplexity project demonstrates the technological recreation of The Bahamas in the name of the Anthropocene.

Although the grand plan to integrate the social, habitat, and genetic connectivity data never came to fruition in a mega model, the project itself was not a failure of socioecologic. The very existence of the project extended the reach of GCS research and marine management strategies in the region. The project represents GCS field research adapted for the small-island and archipelagic condition of The Bahamas and the ongoing expansion of MPAs into marine territory. To this day MPAs continue to spread in the country as managers attempt to reach the national goal of 20 percent of Bahamian marine space enclosed in protected areas by 2020.[32]

Biocomplexity can now be understood as a research trend in GCS that performs its own assumptions to produce new management regimes for the Anthropocene. Far from failing, this project exemplifies GCS socioecologic at work redesigning the governance of human and nonhuman relationships.[33] Through the use of the survey tool, GCS scientists increasingly paint people as variables embedded in the systemic functioning of planetary life itself, in this case in the functioning of marine ecosystems.

I now think of surveys, MPAs, the tourism industry, and even the concept of biocomplexity itself as technologies that have been put to work in re-creating The Bahamas as a new kind of experimental destination. Within the project the survey was designed to record human and marine ecological relationships across island settlements, providing a quantifiable snapshot of variations in environmental perception and interaction between island communities in the country.[34] But the survey reproduced and stabilized popular assumptions about individual and community stakeholders while isolating researchers from research subjects, fixing their roles and the terms of their communication. The survey categories universalized people like Charlie as social archetypes (commercial fishers) as opposed to facilitating meaningful collaboration with people living in surveyed settlements. This technique limits what is possible to learn from respondents, what is possible to discuss, and what is possible for respondents to do as research participants.

If the survey has become a standard tool for demonstrating the social so it can be managed, then the MPA has become a standardized marine spatial management strategy in the Anthropocene.[35] The MPA has been borrowed and adapted from the preservation of the biodiversity paradigm, and it has found new legitimacy within GCS biocomplexity research. The project's social scientists argued that the individuals and communities surveyed have categorical perceptions of MPAs depending on their occupational orientation. But the MPA form itself was not questioned by the project as a management strategy. MPAs and Marine Reserve Networks were instead naturalized as the only resource management strategy that could protect both island livelihoods and marine species, in part because the MPA has been championed by The Bahamas' government as the most viable marine resource management strategy.[36]

The distribution of proposed MPAs, mobile populations of marine species, and Bahamian settlements around the archipelago made The Bahamas an attractive location to test biocomplexity as an emergent research paradigm. The government's MPA expansion plan already implicated adjacent settlements as the appropriate sites for social assessment, and the project assumed that marine managers would target people in settlements close to proposed MPAs. In this way rural settlements become communities assumed to have homogeneous perceptions that can be mediated with education and outreach about MPAs, marine species biology, and sustainable alternative livelihoods to fishing. Part of that mediation stems from the claim that, rather than shutting down livelihoods as fishing grounds are enclosed in "no-take" areas, MPAs are good for the economy because they create new destinations for tourism.

In the years since the project ended, I have come to realize that tourism was an indirect component of the biocomplexity project. Tourism is almost always the elephant in the room when it comes to the enclosure of marine fishing space in management areas in the Caribbean. This is because tourism represents "value added" for environmental management, a value often taken for granted. The association between MPAs and opportunities for tourism and ecotourism ventures is well known, and The Bahamas government and Bahamian environmental nongovernmental organizations promote tourism to compensate for the loss of livelihood associated with proposed no-take MPAs near fishing communities.[37] The rationale is that if there is an MPA in the vicinity, fishers can easily become recreational fishing guides for catch-and-release fly fishing, even if they have never used a fly rod before, or owners of tourist dive operations, even if they have never used scuba equipment before. It is commonly assumed that transitions to tourism-based livelihoods will remove fishing pressure from local waters as a result of MPA regulations and the subsequent reduction in commercial fishing. Therefore, conservation discourse in the country promotes MPAs as good for economic development because of the opportunity to replenish fish stocks and promote tourist visitation at the same time.[38]

The survey utilized by the project implicitly took up this association in its questions by comparing approval of MPAs with employment in the tourism industry, concluding that tourism-dependent communities are more likely to support MPAs. This subtle nexus between social field research and the expansion of MPA-based tourism is almost totally unnoticed in the country because tourism is promoted everywhere. But in this way the project helps perpetuate the sense that the tourism industry is economically stable and ecologically sound. Further, the very material realities of rural lives like Charlie's or T's, lives composed of fishing for love in the crawfish closed season and building traditional skiffs to remain close to the sea in old age, are invisible to the tourism-development desires of the Bahamian government, industry investors, and conservation organizations. These institutions want to see opportunities for tourist visitation as "sustainable" ways to mediate Anthropocene problems like overfishing.

I have spent much more time in both Abaco and Eleuthera since working on the biocomplexity project, but I have been back to Cherokee Sound and Tarpum Bay only once each, years later, passing through both times. The settlements have not changed much in the intervening years. I did not see Charlie or T on those visits, and if anyone remembered me, they didn't let on. I felt uncomfortable returning to these settlements with nothing to show participants after all their time spent answering questions.

The existential challenges of this kind of social project have been articulated by the Bahamian psychologist Niambi Hall-Campbell, who wrote of her experience as a research technician on a social data–collection project within urban neighborhoods in the southern United States in a nice inversion of neocolonial norms. She describes the awkwardness of being a fieldworker, likening it to the relationship tourists in The Bahamas have to the places they visit. Social fieldworkers have the best of intentions, but they have no long-term plan to work in their field sites beyond the horizons of the research at hand, they have few past ties to the site and its residents, and no "sustainable connections" outside the scope of the project. Their actual role is to perform social accountability, leading Hall-Campbell to ask the legitimate question: what do these communities have to show for years of engagement with long-term research projects? The answer is that the researchers tend to benefit far more than the residents in projects like these. Instead of creating good neighbors, these projects built impersonal variables and managerial units out of the social worlds they study.[39]

The GCS scientists for whom I worked were fairly aware of these dynamics, and they recognized that internalizing real people into research projects was not easy. Standardized social categories are seen as a necessary evil, and "social engineering" is part of the "doing of research."[40] Scientists like Dr. Colwell contend that the creation of calculable sociality is actually an ethical stance on the part of researchers because nuance, situated relationships, and anecdote do not appear to solve the extreme problems of the Anthropocene. For these scientists, documenting complexity as it is actually lived does not allow for clear policy solutions like algorithms for siting an optimized network of MPAs.

My experience within this biocomplexity project has shown me that all these technologies attempt to *generate* possible futures. What they don't do is enable any kind of predictability, even if that is their stated purpose.[41] But critiquing projects like this does not mean that we must lose sight of the years of work that have gone into making the natural sciences more socially aware. The shift from the 1980s and 1990s logics of biodiversity to the logics of biocomplexity in the early 2000s was, for some GCS scientists, a major coup. For many social scientists concepts like biocomplexity are groundbreaking in that they allow for the very existence of a social team in a bioscience project. The idea is a kind of victory over a logic that was decidedly asocial, based in notions of pristine nature, which pitted local people as enemies of conservation and wildlife management. But, as I have shown, the possibilities for understanding sociality within the framework of biocomplexity are still deeply constrained.

This project cannot stand for all biocomplexity research or for all science in the Anthropocene. But it reveals that it is now practically logical to design field research projects in ways that consider the human as a key part of larger biocomplex systems. Human life has been exploded into a theoretical cosmology of interconnection, and scientists are experimenting with methods to demonstrate such human connections with far-reaching global processes. The GCS move from the study of diversity to complexity radically reconceptualizes people as connected to other organisms, ecological systems, geologic formations, and biogeochemical processes. But the kinds of connections that GCS can imagine remain stunted.

Ultimately, this project's attempt to refashion The Bahamas into a modelable socioecological system reveals the tensions within Anthropocene science. GCS acknowledges that human activities are a driving force in planetary ecology, and biocomplexity research is indeed more social than most ecological field science. But after more than twenty years of theoretical development, this is still a reductive sociality. The universal social categories of biocomplexity limit what can count as variables, proscribing the subjects and objects of research. They make targets for research and policy intervention out of community members, but they do not enable communities to speak for themselves. Simultaneously, biocomplexity's limited scope produces potentially lucrative social targets for GCS projects, environmental governance, and the tourism industry, without necessarily considering the effects of substituting fishing for tourism within expanding MPAs. The survey data can explain what it costs to fish, how frequently individuals go fishing, and how likely a community is to approve of a local MPA, but it can say almost nothing about what it means for Bahamians like Charlie and T to live in the Anthropocene Islands and what else they might want for the future.

2

The Educational Islands

AN ISLAND EDUCATION

In 1648 a group of Puritan religious dissidents left Bermuda and sailed to seek their fortune in The Bahamas. Bahamian history has it that they endured a perilous journey to the island known to the indigenous population as Cigateo, renaming it Eleuthera, the Island of Freedom. This small group, now known as the Eleutheran Adventurers, was the first European effort to formally colonize The Bahamas. The arrival of Columbus to The Bahamas in 1492 is possibly the most (in)famous story of arrival, but the Eleutheran Adventurers legend is a well-known settler story within the country today, and many Bahamians trace their lineage back to these first English settlers.[1] The Spanish were said to have killed and enslaved the indigenous population of the islands many years before, and the Adventurers' arrival in Eleuthera begun the transformation of the supposedly "empty islands" into a place of civilization. The *terra nullius* narratives of this group set the tone for present-day Eleutheran adventures in the realm of education, space making, and designs for living in the Anthropocene Islands.

In The Bahamas tourism and its infrastructure represent a contemporary attempt at settlement as much as development. The Bahamian tourism model of the late twentieth century was the resort enclave, carefully situated in a prime location on a beautiful island as an economic "anchor." The resort was a corporate outpost of the tourist metropole, and lucky guests were flown or shipped in to their destination and flown or shipped

FIGURE 5. Street sign for an unfinished development, South Eleuthera. Photo by author (2008).

out again after a few days of all-inclusive fun. This, along with cruise touring, is still the dominant form of tourism in The Bahamas.

The island of Eleuthera has had its share of experience with the anchor model of resort development. South Eleuthera, especially, has seen several large enclave resorts come and go in the past sixty years, precipitating the familiar cycle of boom and bust in the Family Islands.[2] The southern end of the long and skinny island is dotted here and there with developments in ruins. It was in this economically depressed environment, this "reemptied" island, that the Island Academy was developed.

The Island Academy is a U.S. nonprofit institution for experiential high school education located in South Eleuthera. The school was built from the ground up among the bones of an old resort abandoned years ago. On one visit to the academy I wound up bicycling down overgrown roads on a rambling ride to nowhere in particular. My guide on that day was Douglass, the only Bahamian member of the academy's teaching staff at that time. Douglass was in his late twenties and enthused about finding a job in The Bahamas after spending some years in the United States getting a degree in fine art. He wanted to show me the landscape as a way to understand the school's proximity to island history. As we rode, I marveled at the abandoned roads and the signs that announced themselves on every vacant corner, identifying lanes and streets that no longer held the possibility of foreign investment. Only mockingbirds, flycatchers, and butterflies populated the dry bush along the roadways. On this cycling expedition I saw that evidence of the collapse of this twentieth-century island travel mecca is literally carved into the limestone. Overgrown driveways lead to rotting bungalows; the sprawling plan of an old golf course can be identified under the dense tangle of bush; weedy airstrips provide opportunities for drag racing or bicycle speedways for children.

The academy is another example of Anthropocene logics at work in The Bahamas. In this case one specific island, Eleuthera, has been rebranded as a mecca for education: a model system for visiting students, researchers, and investors. Focusing on the Island Academy, the Island Institute, and the now defunct offshoot, Island Solutions, reveals the ways in which private experiential education and scientific enterprise attempt to market this "educational island" and produce future-oriented Anthropocene student subjects.[3] The Island Academy represents one way that private capital has attempted to turn the crises of islands in the Anthropocene into "sustainable solutions" and a form of targeted travel opportunity.

INCULCATING THE ANTHROPOCENE ISLANDS

The national educational system of The Bahamas is run by the Ministry of Education, and each island in the Commonwealth of The Bahamas consists of a school district or a number of districts, depending on the island's population. Each district provides primary and high school education. In 2010 the Eleuthera and Current Island District had approximately 2,500 Bahamian students, eleven primary schools, three high schools, two all-age schools, and a teaching staff of approximately 180, most of whom were women. Most Bahamians attend public school through high school, at which time many enter the full-time workforce, though this trend has been changing for some time. For those able to continue their education, some attend the University of The Bahamas in Nassau on the island of New Providence, while the elite are sent, and some of the best students obtain scholarships, to foreign universities in the United States, United Kingdom, and Canada.[4] The Ministry of Education creates national curricula for all subjects, grades, and ages, and these curricula are taught in all the public schools.[5]

The islands of The Bahamas are also home to a number of scientific research stations, created by foreign and domestic organizations for the education of foreign and local students alike, as well as for the development of field research within the country. The list of research stations

FIGURE 6. High school girls waiting for a jitney after school, Nassau. Photo by author (2008).

has included the Bahamas Environmental Research Center on the island of Andros, the Gerace Research Center on the island of San Salvador, and the Perry Institute for Marine Science in the Exuma Cays.[6] The Island Academy, created in Eleuthera in 1999 by teachers from a private U.S. high school, is one of these research stations.

The Island Academy is an educational semester-abroad program for primarily white and upper-class U.S. and Canadian students who, over the years, have come from at least 250 schools. A few Bahamian students have also attended the academy. The campus is located on the southern half of Eleuthera, just east of the central island of New Providence. Students pay to live in small dormitory-style buildings and take "place-based" field classes in the Bahamian marine, terrestrial, and social environments, including reef studies, aquaculture, waste management, social studies, and archaeology.[7] Island Academy courses are not connected to the Ministry of Education's curricula, and the academy designs and maintains its own classes. For many years students also participated in a one- or two-night home-stay program with residents of a nearby settlement. These programs are designed to "connect students to place." The academy faculty are almost all U.S. college graduates, again primarily white and upper class, with a higher percentage of male teachers than the Bahamian system. The cooking and maintenance staff are mostly black Bahamian, Haitian, or Jamaican from the rural Eleutheran area. These asymmetries of race, class, and nationality are not at all exceptional when it comes to postcolonial field schools anywhere, but what is exceptional here are the explicit ways that the Island Academy has turned its island location—island people included—into a product for the Anthropocene.

I have visited the Island Academy on a number of occasions. My first visit was in the summer of 2005 and my last was in the spring of 2014. In 2008 I was invited to stay for an entire week at the behest of Douglass, who was also the Bahamian outreach coordinator in addition to being an instructor. He thought my research was timely, and he wanted me to speak to students at the school as part of his responsibility for scheduling their continuing rotation of guest speakers. I flew by small jet from Nassau, arriving at the tiny airport in South Eleuthera early on a Monday morning, when the academy's day was already well underway.

Any visitor to the Island Academy is struck by how regimented the student schedules are, extending far beyond the bounds of a typical school day. Some people attribute this discipline to the fact that the head of the academy was once a Navy Seal within the U.S. military. But whatever the intended reason for this highly organized schedule, one result of this labor is the inculcation of a particular sort of subject position in students. The school day begins with collective morning exercise at six thirty for the students and faculty, followed by chores (cleaning dormitories, tending the compost, feeding livestock); communal breakfast; guest speakers or project research; classes including math, art, or the theme of the week; communal lunch; more classes; scuba or community outreach or advisory meetings; one hour and forty-five minutes of free time; communal dinner; then study time until bed around ten o'clock.

In other words, the schedule is busy, structured for nearly the entire day, and focused on the campus or areas nearby. The routine is physically palpable, becoming ingrained within a few days. Every class is tied directly to an environmental lesson in or about the local surroundings. Repeatedly, in classes and at group sessions, students are asked to discuss how their island education informs their personal awareness of the need for environmental conservation. And this disciplining in island-based education is not limited to formal instruction hours. The material infrastructure of the school itself, and the island of Eleuthera, is utilized by design to inculcate an appreciation for island vulnerability and sustainability.

On the first day of my visit I was given the tour of the facilities, which cover several acres of arid and rocky land by the sea. This tour is what sells the Island Academy and its plans, and visitors today would be given the same tour. My guide on that day was again Douglass, and he confided in me then that he had come to work at the academy for a season or two in part because he wanted to see what sustainability science could do for the country. But, as one of the only Bahamians on staff, he was personally in charge of presenting information about present-day Bahamian people to the students and for inviting people from other parts of the country to come talk about The Bahamas.

The first thing Douglass showed me was the waste management system, of which everyone was very proud. The academy's septic tank has been linked to a simple biodigestion system, wherein human waste feeds the plants displayed in the center of the campus through an underground filtration mechanism. The tropical garden was tall and lush when I visited, though Douglass noted that one could occasionally catch a whiff of the "fertilizer" from time to time. I was told that this was only a small, low-maintenance example of the academy's waste management plans and that they hoped to design ways of turning waste into energy through the extraction of methane gas for cooking and electricity, even to extract water from it.

They say that it is all quite simple: the academy has an agreement with The Bahamas Electricity Corporation, wherein it can produce its own solar power and connect to the island's public electricity grid through an intertie system that allows it to pump power into the grid during the day and pull power off the grid at night when the sun is not shining. This has come to work so well, I was told, that the academy actually stabilizes energy in the area. The academy produces its own water from water cisterns collecting rainwater, though it has a back-up well field for times of drought. They do not purchase water from the government, unlike most people on the island and in the country who can either purchase drinking water from private corporations or who are provided with household water as a public utility. They even get their internet and phone services by satellite from New Jersey, avoiding dealings with the Bahamas Telecommunications Corporation altogether. The academy must import most of its food, however, though they grow their own lettuce and occasionally slaughter a pig on parents' weekend as a demonstration for students on the hidden processes of consumption.[8] The whole campus is explicitly designed to become as self-sufficient as possible.

In addition to the infrastructure, there are other designs for the student's hands-on education and for the purpose of demonstrating "the possibilities of what can be done" in an island setting. I was shown an aquaculture site wherein students and interns maintain and propagate populations of edible fish, such as tilapia. In an adjacent aquaponics area, students learn how the waste from the aquaculture fish tanks can fertilize the growth of hydroponic vegetables, vegetables that purify the water, water that is then funneled back into the fish tanks. Much of the lettuce eaten at the academy is grown in this manner. There is a plant nursery in which both native and nonnative plants are grown for landscaping, and there is an orchard irrigated from the academy's water supply. There are large composting bins, divided into various categories and mode of decomposition, and some of the compost goes to feed the large hogs in the academy pig farm. The campus furniture is made in a woodshop from the wood of the Casuarina, an invasive pine that can be found in abundance throughout The Bahamas. The facilities showcase a number of green buildings, one with a living roof and a few with earth-bag construction in lieu of cement. Seaweed is collected and used as additional plant fertilizer. There is a wind generator that can produce up to fifteen kilowatts. Lastly, there is a small biodiesel-production shed wherein cruise-ship cooking oil is transformed into diesel for campus vans and machines through a simple procedure involving the separation of glycerin. This fuel is sold to faculty and staff at a price that is consistently at least a dollar less than the price of gasoline on the island.[9]

These site-based demonstration projects combine with the academy's science and humanities curriculum to create a strong and unique sense of place on the island for students. The sustainable systems provide the basis for an "intentional community" at the academy, designed to cultivate a strong student "sense of self." And whether based in literature

FIGURE 7. The Island Academy wind turbine, Eleuthera. Photo by author (2014).

and writing or applied mathematics, each class has a field component that must be applied to what the academy sees as the "real world problem" of "living better in a place."

In terms of content the science courses are linked to a number of ongoing research projects that use the marine and terrestrial ecology of the island in various ways. Every day the students participate in projects designed by the Island Institute, an adjacent and affiliated center for scientific research and visiting science professionals, completing various tasks and working in small groups. When I was visiting, the island-based student research projects included aquaponics for sustainable food production, archaeological analysis and GIS mapping of historical Lucayan sites, coral reef ecology and the recovery of long-spined sea urchins, offshore aquaculture with cobia fish for potential commercial production, patch reef monitoring, and local shark-population assessments.

The humanities courses involve frequent trips to nearby Bahamian settlements, where students conduct "mini-ethnographies" that provide "cultural immersion experiences" with local residents to learn from "people with a past." Students are also asked to participate in "stakeholder meetings," where they attempt to explore how Eleutherans can collectively "make change" by identifying problems with Eleutheran socioecological systems such as waste management or fishing practices and examining islander approaches to these problems in the meeting to understand "stakeholder perspective and bias."

I visited the Island Academy in the middle of Tourism and Development Week for the high school students. Activities were geared toward explaining this topic in a place-based way. For example, I accompanied students in field trips to two nearby resort developments, Cotton Bay and Cape Eleuthera Resort, where we were given tours of the premises and two separate sales pitches from the property managers.[10] I also sat in on several guest-speaker presentations by members of the Eleutheran community. Before each short presentation students were admonished by faculty to think of the speakers from the island as a lesson in "living history."

At the end of each semester season at that time, students presented their work in a research symposium in an attempt to show island leaders and representatives from national environmental organizations—the aforementioned stakeholders—that they have identified unsustainable problems with island life and devised sustainable solutions. The curriculum is therefore designed to link content based on a particular vision of sustainability and human ecology to place-based field experiences, public speaking, and leadership to provide a comprehensive "transformative experience." When students leave the Island Academy, they are supposed to be forever changed.

In 2008 I could not interview any high school students directly.[11] But I was fortunate enough to meet Lamar on that visit, then an eighteen-year-old Bahamian intern at the academy. Lamar is from Nassau, raised in a middle-class family with two parents who work for the government in civil service. Standing out from the majority of academy students, he is very tall with dark skin, long dreadlocks, and a giant smile. He was excited about being interviewed by a visiting scholar, and he answered my questions with enthusiasm. He explained that he had spent a semester at the academy right after he finished public high school in The Bahamas at the suggestion of his marine science teacher. He had subsequently returned to Eleuthera the following summer to be a paid academy intern, receiving a small monthly stipend plus room and board—less than what he would have made working in Nassau in the tourism industry but enough to support himself in Eleuthera with virtually no expenses. He thought that the academy was a good educational experience, encouraging his interest in science and ecology while helping him feel comfortable about public speaking.

During our conversation Lamar confessed that although he was interested in everything the academy and the institute devised, he was especially interested in aquaculture. His internship responsibilities revolved around an offshore experiment with cobia, and he had dreams of promoting cobia aquaculture as a business model throughout the islands of The Bahamas in the future. He confidently stated, "As people see your success, they are likely to start their own business." On the way to becoming an eco-entrepreneur, he hoped to study biochemistry and marine biology in college. He realized he would likely have to start at the University of The Bahamas, but he told me that he might be able to go away for a graduate degree or a second undergraduate degree at some point afterward.

Lamar had also struggled a bit at the academy. He became homesick during the solo kayak trip every student has to make, sleeping out alone on a beach for two days with only a small bag of food and water for company. He also found it awkward to do the one-night homestay in the nearby settlement with a Bahamian family, considering he was also a Bahamian. He told me that he had primarily answered that family's questions about life in Nassau and his parent's occupations. He wasn't yet accustomed to the freshwater rationing that was a part of life at the academy, stating, "You can't just use water there as you would at home." In addition, Lamar though that there should be more Bahamian staff and advisers like Douglass at the academy, though he admitted that this was a tall order, considering the remoteness of the school and the "style of conservation" practiced there that permeated every aspect of daily life.

Despite these challenges, overall, Lamar found the academy experience to be overwhelmingly positive. He enjoyed his time at the academy primarily because of his social life. He now had friends from all over the United States. These student friends at the academy were very interested in him because he was Bahamian, and they asked him all about what he ate, how he lived, and how he spoke. He even taught a class on Bahamian slang for the students, "just for fun." For the other students Lamar's form of living history was youthful, exuberant, and humorous.

I provide all this information here to demonstrate only some of the ways in which the academy differs radically from the style of public education offered by the Ministry of Education. The private Island Academy has come to use its island setting and highly orchestrated student activities to maximum advantage as elite education. Through the disciplining process of experiential education for student visitors, the school adheres to its mission to be "much more than a place of learning. . . . Students are active participants in the educational process; students have to think like scientists, cultural historians, and teachers. . . . They face real problems and challenges in and out of class."[12] The Island Academy has turned its very infrastructure, as well as Eleuthera's ecological systems, people, and its conservation and development challenges, into an island classroom for the inculcation of designs for life and living and the production of student subjectivities oriented for the Anthropocene.

SCIENCE TOURISM IN THE EDUCATIONAL ISLANDS

Beyond the regional notion of the North American student as the "spring breaker," the Island Academy is an example of a situation that I call the "educational islands": tourism designed around teaching and learning for the Anthropocene with a particular target—the malleable and affluent student inculcated to be a potential future leader, global change science (GCS) scientist, teacher, or foreign investor. A representative from the Ministry of Education who lives in Eleuthera noted in my presence that the academy kids might be visiting students now, but they might also be potential investors in Bahamian development later and should be *treated as such*. This Bahamian official was expressing the general position held by many South Eleutherans toward visitors they perceive to be tourists: we want them to come because we want them to build space for more. We want them to pay us for our labor, but we don't expect to interact with them much beyond that. That is how anchor resort enclave development is supposed to work.

For South Eleutherans the academy was built on the site of a former failed resort without much fanfare or notice. Few people knew what it was until students were already there. As the academy grew and its plans to make the island a holistic place of learning expanded, students, staff, and island people increasingly met and sometimes clashed. The academy's self-sufficiency was perceived by some as a snub to South Eleuthera and its workforce, to people who barely maintained a living from what few opportunities there were already—mostly civil service, fishing, and very small businesses. As the school developed what is seen as its top-down management agenda, pushing publicly for fishing regulations and the creation of a nearby marine protected area, some island residents began to see the academy as a nuisance or, at worst, as an insult to the style of anchor tourism synonymous with The Bahamas. The academy was not fulfilling its part of that tourism-development bargain.

The academy responded to this criticism in 2001 by opening an academy-staffed middle school for Bahamian students in the closest Bahamian settlement, by admitting a few Bahamian high school students on scholarship every year, by inviting local people to its annual research symposium, and by always offering to give anyone the tour. The results to date of these efforts are mixed, and many people who live nearby have still never visited the academy. Yet for Bahamians like Douglass, born in Nassau, educated in the United States and the United Kingdom, and also a visitor to Eleuthera, the school holds out a different kind of promise. For Douglass the academy is indeed a school that must attract foreign students who can pay their way and support its plans on the island, but it is also a quasi-utopian model for sustainable island dreams.

The Island Academy is a site where GCS field research and island ecological education implicitly mix with the tourism industry of The Bahamas. Target students are potential foreign investors as much as future leaders. But this is not an openly discussed aspect of the educational islands within the academy itself, and it is one of the few aspects of students' lives that does not make it into the syllabi. For over a decade students and faculty from North America have traveled to Eleuthera in small numbers, usually with no prior experience of The Bahamas. They go there to be a part of the academy's mission to create "a community that fosters the development of responsible, caring global citizens by restoring a sense of wonder and respect for biotic and cultural complexity."[13] They go to experience life on a remote (though increasingly less so) Bahamian island, to immerse themselves in the cool shallow sea and dry heat of the bush, and to have a unique sort of adventure that promises to improve their understandings of themselves, their fellow adventurers, and the fate of the changing globe through science-based field education.

The academy imagines that both the planet and student subjectivity are malleable, improvable, and receptive to positive change based on knowledge. This is a form of "science tourism," an underdeveloped arena of scholarship, and it is as individualized through the development of the "student self" as it is globally conscious. Further, the location for the adventure, the island of Eleuthera, with its bush, mangrove, reef, and beach ecosystems and particular social history, is instrumentalized here in curious ways. Through the spatial framing and practice of the field studies and projects, students learn that the Bahamian landscape is composed of local people, animal species, marine and terrestrial ecosystems, and processes of natural and anthropogenic dynamic change that cause the availability of natural resources to be uneven. They learn that GCS science is the language with which to speak truth to the uninformed to influence the creation of "good" policy decisions, and they are charged with demonstrating the scientific validity and social relevance of their projects.[14]

The study of science tourism remains underdeveloped.[15] For some scholars, however, scientists can be considered visitors because they benefit the tourism industry with their money, their return trips, and travel recommendations to friends and family. They are considered early explorers of areas that may become more widely visitable over time. This form of tourism is relevant to small countries like The Bahamas because of their desire to attract special interest segments of the global travel market.

In light of these points, it is clear that the Island Academy, and the students and faculty it attracts, is very much an active part of the tourism industry of The Bahamas, as are all visiting researchers, myself included.[16] While research centers complicate things in that they often funnel earned money out of the country, back to the center's place of origin (for the academy, this is the United States), this is not much different from the dominant form of foreign-owned and -operated resort tourism that is so widespread in the country and the region. There are a great many forms of science tourism, from single researchers who camp alone in their field site to situations like the Island Academy in which students and faculty travel en masse to live and work at a center in Eleuthera.

A "tourist" has been defined in anthropology as "a temporarily leisured person who voluntarily visits a place away from home for the purpose of experiencing a change."[17] While the students, faculty, staff, and interns at the Island Academy are employed in educational activities, they can still be said to be traveling for the purpose of experiencing a change. In this case change comes through the act of participating in scientific knowledge production wherein the research comes to justify the experience. But, unlike critiques of "tourist products" that focus on the commodification, Westernization, or Americanization of social forms in tourist markets, what is at stake in southern Eleuthera is not necessarily the loss of culture for Bahamians in the vicinity, nor is it necessarily that people are forced to commodify their lifeways for tourist consumption, although this does indeed occur in some cases.[18]

Instead, the stakes at the Island Academy involve the production of visitable anthropogenic complexity itself as a scientific tourism product, through the generation of a tightly constructed island experience. They concern the development of new forms of situational education and the ways island dwellers and developers conceive of living spaces, infrastructures, and modes of self-awareness. This takes on even more pointed meaning when representatives from the academy describe planet Earth itself as an island.

ELEUTHERA 2030

The Island Academy is a branch of the Island Foundation, created in 1996 by a wealthy U.S. family with investment interests in The Bahamas. The foundation is an umbrella organization, supporting several ventures, including Island Solutions and the Island Institute. Island Solutions was created in 2005 to develop "sustainable alternative energy industries" for profitable application in island settings.[19] This was considered a natural extension of the work done at the academy, and Island Solutions became the business end of the school as the revenue-generating arm of the foundation's entrepreneurial model. The website notes, "As a wholly owned subsidiary of a not-for-profit corporation, [Island Solutions] is able to provide support to environmental and social projects through the [Island Academy] that are aligned with its ongoing missions of education, research, and outreach. In this way we are creating a model of applied business practice, rooted in sustainability, and tied to the education of future leaders. [Island Solutions] is building these models on the foundations of evaluation, quality service, and cutting-edge technology." In other words, for the designs developed at the academy to be considered truly sustainable, they had to be shown to be profitable.

The scientific practices that transform Eleuthera into an educational island are explicitly linked with enterprise, and the most visible enterprise, the one most missed in the area, is tourism. In 2008 Island Solutions proposed *Eleuthera 2030,* their most ambitious design to date. It was a blueprint for the proposed future development of Eleuthera by the year 2030, with Island Solutions situated as the prime mover. The full title for the proposal was, *Eleuthera 2030: An Island to Reinvent the World,* and it was touted as "a tremendous opportunity to make a real impact in island energy systems and brand both Eleuthera and The Bahamas as a cutting-edge destination that is truly engaged in sustainability."

The "opportunity" imagined here stemmed from the inefficiency and expense of Eleuthera's island energy system and from the fact that Eleuthera is small enough to be managed as a whole. In their view the island was ripe for redesign and rebranding. Eleuthera's standard means of producing energy comes through the consumption of imported oil and gas resources from other nations, yet the proposal pointed out that the Caribbean's main resources are the sun, the wind, and the ocean. Island Solutions argued that changing the means of powering the island would affect the cost of energy, the potential for carbon taxation, vulnerabilities to climate change, the tourism industry, and quality of life for island residents. Knowing that there were only 8,300 people on the island, not counting tourists, with known electrical and gas consumption and that the island has a solar resource in its clear, sunny weather, Island Solutions proposed that it would take only a small surface area of solar panels to power the whole island. They noted that wind turbines could be placed on ridges or in the shallow sea. The proposal sited the Island Academy as its best example of success, noting that it generated 60 percent of its own energy and fed energy back into the Bahamas Electricity Corporation grid. They even proposed the development of a large biofuel plant to produce jobs and generate carbon credits for The Bahamas. There would of course be large expenditures of money required up front to fund this vision, but the *Eleuthera 2030* argument was that these were low-maintenance systems with few costs after setup.

Even though it has nothing to do with industries like genomics, the *Eleuthera 2030* venture is meaningful in part because it mirrors developments in biotechnology. In both the fields of biotech and sustainable living systems, facts produced by science about nature are linked to the speculative tendencies of contemporary economic markets. In a sense life has become a kind of "business plan" that can be invested in. Nature, therefore, is scientifically produced as "life as form of fact" and articulated with processes of valuation through a process that some scholars call "venture science." In this case it is the research that determines the shape of the product, defining the ways in which life can be internalized in market calculations.[20]

The *Eleuthera 2030* proposal devised by Island Solutions is therefore an example of island life scientifically reproduced as an Anthropocene product. Under the sign of sustainability, with its explicit future orientation, Eleuthera is remade into "life as form of fact," and its ecological and social systems are studied in ways that allow them to be invested in for potential profit. *Eleuthera 2030* is not just a proposal for a more efficient way to power the island—it is explicitly a business model, proposed to entice investment from the Bahamian government and private enterprise in island-appropriate alternative technologies maintained by Island Solutions. Even the science tourism of the Island Academy itself is another model for enterprise, enhanced by the brand power of Island Solutions, though this time it is the parents of prospective students and potential donors being asked to invest.

In the end Island Solutions never caught on with many investors outside of the Island Foundation, eventually rebranding as a sustainable internship program for U.S. college graduates. But within its Anthropocenic designs, the same geographic and demographic features of the island itself that are still utilized to make an educational experience for student visitors at the Island Academy were utilized by Island Solutions to promote the creation of a "cutting edge model" and business proposal. Photos of tanned and smiling students enjoying the sun and sea on the Island Academy website gave way to green-and-blue satellite images of the entire planet, then The Bahamas, then the island itself, in the *Eleuthera 2030* proposal.

CARIBBEAN ISLAND LABORATORIES

While at the academy, I managed to have several conversations with staff members during rare moments of downtime. They were generally pleased with the rigor and organization of the school, but some were critical of the style of teaching. One American who holds an important long-term position on the academy's administration went so far as to tell me, "There is nothing neocolonial about the Island Academy—it is a straight-up colonial model." They went on to say that students are never asked to reflect on their own positions as visitors and consumers in the country and that they are more often focused on spreading a conservation and sustainable development ethos with a missionary zeal. Within all the activities all the sustainable living systems design, all the tight scheduling—within all the labor that goes into inculcating this ethos for students and visitors—there appears to be no room for a reflexive acknowledgment of the academy's complicit role in the historical patterns of settlement in the country or of its relationship to tourism and foreign investment.

SCHEDULE FOR MONDAY, MAY 26, 2008

6:30 A.M. Morning exercise

7:30 A.M. Chores

8:15 A.M. Breakfast

9:15 A.M. Local guest speaker

10:15 A.M. Math/art

11:15 A.M. Math/local guest speaker

12:15 P.M. Lunch

1:15 P.M. Math/local guest speaker

2:15 P.M. Art/math

3:15 P.M. Community outreach (designing promotional marine protected area posters)*

4:15 P.M. Free time

6:00 P.M. Dinner

7:15 P.M. Study hours

10:00 P.M. Bed

*Two days a week this slot also becomes designated scuba time.[21]

Internal criticisms of the academy situate the discussion of educational islands in the Anthropocene within the larger perpetuation of colonial problems in the postcolonial present.[22] The characterization of the Island Academy and its partner organizations as a "colonial model" is an important frame with which to consider research stations, science tourism, and the reformation of island space and subjectivity in the region. Colonial and postcolonial questions of race, class, identity, power, and social change remain essential lines of conflict in the Caribbean and throughout much of the world, and at the same time efforts to intervene in local realities under the sign of sustainability are increasingly common throughout postcolonial nations. Therefore, traveling

GCS researchers must consider the institutions they work with and the colonial themes they raise, no matter where in the world they work.

Considering the history of The Bahamas, a former British colony, and the Caribbean region, an exemplary site of European colonial intervention and violent exploitation, the phrase "colonial model" is quite loaded.[23] The staff member who described the Island Academy as a colonial model did not elaborate on what was personally meant by that designation, but based on my familiarity with the site I can reasonably conclude that the referral was to a hierarchical and racialized structure of dominance and dependence in which visiting researchers, students, and faculty (primarily upper class, Euro-American, and white) assume that they have an authority over knowledge production that rural Bahamians (primarily lower class, island bound, and black) do not. The general attitude among students and staff is that rural island Bahamians lack the appropriate scientific information with which to make responsible decisions.

The postcolonial present is a point of departure, an opportunity to engage with the specific conditions of the colonial era in ways that consider the powerful productivity of the past and present moments together.[24] For many scholars the postcolonial is not a transcendent era, but one in which particular colonial relations exist in the present. The staff member was making a pointed statement about the role of the Island Academy in perpetuating historical inequities of knowledge production in the region because the academy privileges what could be called "Western" or "Euro-American" technoscience tied to Euro-American experts and capital. The academy links its technoscience to a moral argument about the regional dissemination of this knowledge in the form of mandatory "outreach," a practice that is especially problematic when the moral imperative is tied to the desire to make outreach profitable.

The staff member's statement about colonial models highlights the powerful creativity of the projects, designs, and ideas stemming from field research stations. For example, the Island Academy's efforts to make Eleuthera into an experimental space for island-based field study revisit themes from "classic" Caribbean scholarship, conducted in colonial times.[25] Social science concepts such as acculturation, transculturation, and creolization were developed in the Caribbean on the islands of Haiti, Cuba, and Jamaica, making each island into a "laboratory situation" for anthropological notions of social change.[26] Anthropologists used their Caribbean island field sites to produce ideas about a more general human nature based in regional history, its capacity for change, and the specific social processes that result from this nature over time. These concepts were explicitly political, with islands situated as experimental arenas for the refutation of prior claims about racial inferiority and as locations for the development of potential national and ideological futures. To the island laboratories of "classic" Caribbean scholarship, we can now add the designs of the Island Academy and the formation of Eleuthera as a laboratory situation for the Anthropocene Islands created through processes of visitation and settlement.

Considering the ways in which colonial Caribbean islands were used by social scientists to produce new concepts about human cultural contact makes it difficult to summarize the Island Academy as a "colonial model" in only a narrow oppressive sense. That perspective can only partially explain the designs produced at the educational center. Instead, the Caribbean examples point to the ambivalence of a Caribbean island GCS research station in the Anthropocene. I do not mean to downplay the perpetuation of inequities between visiting staff and students and Eleutheran residents or the rather narrow-minded assumption that the academy ought to be the foundation for Eleuthera's salvation. Nor do I mean to downplay the feelings that the staff member might have had in speaking to me about a personal take on the academy. These observations led to important criticisms that the academy must entertain if it is ever to seriously come close to fulfilling a mandate to serve the island, but these examples also open the door to the creativity of the educational islands.

We must remember that the phrase "colonial model" also refers to the great capacity of colonial institutions in The Bahamas and the Caribbean to remake islands and island populations into resources while delimiting the possibilities for island life.[27] The trope of educational islands for living in the Anthropocene extends this pervasive creativity, making a living laboratory out of The Bahamas. But at the same time we must not forget that this is a tense and uneven creativity. Within the practices of the Island Academy, Eleuthera is not produced as an educational island for everyone in the same way. Bahamians attending the academy are not visitors to an island experience. They are conscripted subjects, and they are not considered to be central actors in the entrepreneurial schemes of the school.

AN ISLAND TO REINVENT THE WORLD

In 2014 I reconnected with Douglass on a visit to Nassau. Now in his late thirties, he was no longer working for the Island Academy, having left some years earlier, abandoning his bicycle and Island Academy rental housing. He was enjoying being a smart, young professional in his hometown. I asked him how he felt about the academy and the institute now that some time had passed since he last worked in Eleuthera. Douglass was reflective. He said confidently, "I think it needs to be duplicated as a different tourism model in other islands." Rather than describing it as a research center, he saw the academy and its offshoots clearly as a development model—a tourist destination that is also a hub for sustainable innovation—that might still work forty years into the future. "What else is there?" he asked me.

Douglass had been so impressed with his time at the academy that he flirted with replicating this model himself. In 2015, on yet another of my many visits to Nassau, he told me that he was weighing the prospects of opening a hostel for young travelers that would feature a number of green design features adapted to a remote setting in the Family Islands. In other words, he was considering creating his own sustainable destination for affluent but frugal, educated, and connected professionals or students. But by 2016 his attitude had changed slightly.

On that next visit Douglass confessed to me over coffee that he had left the academy because he had eventually stopped following the daily routine, a form of passive resistance to the school's disciplining gravitational pull he described as "wearing him down." Douglass explained that he had spent time teaching and advising the U.S. students, and he saw firsthand how wealthy some of them were. He told me about a few that even arrived to Eleuthera on family-owned jet planes. Douglass was well aware, along with the rest of the staff, that these students were potential future investors in Bahamian property and in the academy itself, and he characterized the institution as a school for "rich American kids." Although he supported the fact that Bahamian students were welcome there if they successfully applied and were eligible for scholarships, he knew that the academy was not designed to be a school for most Bahamians.

Douglass went on to say that the academy had a lot of influence in the country, but not so much that they could tell the government what to do. For example, he recounted a bid to the government, based on the *Eleuthera 2030* vision, that Island Solutions had made to design alternative energy systems for the country as a whole. Once the government made it clear that it would not hire Island Solutions, "they didn't push it." In other words, they had encountered the limitations of their influence as a foreign nongovernmental institution in a sovereign postcolonial context. Yet, Douglass noted, one parent wrote them a check for $1 million for another project. A lot of this kind of parent-sourced money went into funding academy projects. He was impressed at the ease by which this organization could access this form of private funding.

In the end Douglass conceded that the visiting U.S. students certainly had transformative experiences, and it was a valuable time in their lives. In 2016 he no longer wanted "anything to really do with the place," but he stressed to me that the academy had influenced him to think about sustainable policies for The Bahamas and to strategize about how he could be involved in those plans as an entrepreneur, even though he did not have his own funds to start a business without outside investment. He had to give them credit for being innovative in the way they mix tourism with enterprise.

I also reconnected with Lamar on that 2016 visit to Nassau. He was now twenty-six, still a giant with a warm smile, and happy to meet with me to talk about his life. He had graduated from the University of The Bahamas with a major in biology and minor in chemistry and was glad that he hadn't gone to the United States for college because "people who school away don't come back." His Island Academy experience fueled a continuing love of coastal ecology, and he had since participated in a number of internships and volunteer opportunities with visiting researchers and citizen science organizations. He even completed two more summer assistantships at the academy in 2014 and 2015, working in their lionfish program, tagging bonefish, and exploring inland ponds. He had become a dive master. When I caught up with him, he was working at the Bahamas National Trust but planning to attend a Caribbean regional university outside of The Bahamas for a graduate program in coastal management that coming fall. And, unlike other Bahamians he knew who did not return after studying abroad, he was convinced that he would return to Nassau with his degree.

For the second time in his life, Lamar described his plans for the future while I diligently recorded them in my notebook. After finishing his degree he hoped to work for an environmental consulting firm in The Bahamas with the end goal of owning his own firm once he had amassed the capital to do so. He wanted to use this firm to partner with other universities and engage young Bahamians in research projects and assessments for paying clients. Lamar explained that because of his experiences with the Island Academy, he would rather be a "change agent" for private development projects than an educator in the formal school system. This was based on his perception of the country's debt and reliance on foreign investment as a small-island state. He said, "We are heading to a lot of big problems. A meltdown is coming soon, and then change will happen drastically—a real bad crisis. It will inspire everyone to change their ways," and everyone in the country will need to invest in a sustainable small-island economy.

Eleuthera's "islandness" is a large part of what makes it an exemplary place for research, experimentation, entrepreneurialism, and redemption in the Anthropocene. This islandness is what the Island Academy and Island Solutions hope to capitalize on in their various ways. The island is a curious unit, connoting the interplay of separation and interrelation. The scientific productions of the academy, as an educational center, perform the vulnerability of island systems, the geographically determined locality of island organisms and communities, and the accessibility of the field for those who have the right kind of knowledge. The island is a product. It is where one arrives as a visitor to learn how to manage and imagine larger global systems and where the tourist becomes transformed into a global leader through a labor-intensive conscripting process involving the orchestration of space. The productions of Island Solutions, as a model for enterprise, attempt to demonstrate that the challenges of the island are the challenges of the world, that the planet is itself an island, and that the appropriately redesigned island, informed by science, can become our common future in an era of anthropogenic planetary change. The island is an iconic Anthropocene spatial product: a visitable site of ecological, social, and economic salvation made recognizable through GCS in the Anthropocene.

Through the productions of the academy, Eleuthera becomes a reinvented colonial model: a living laboratory for the development of scientific and touristic enterprise. As a revisitation of the Eleutheran Adventurers, the scientists, students, and teachers act as "early explorers" who will open up the "empty" Bahamian frontier to new forms of scientific arrival and responsible intervention.[28] What is produced here by the branches of the Island Foundation is both a business model for the Anthropocene, promoting the "triple bottom line" of economic, environmental, and social development, and a concomitant socioecologic for redesigning living systems.

The Island Academy demonstrates that we must take the imbrications of science seriously. Anthropologists note that ecology is a quintessentially "worldly science," and scientists have long been involved in solving practical problems along with extractive industries.[29] This practicality has led to the assumption in the Anthropocene that extractive interests, including tourism, have the most at stake when it comes to protecting ecological resources. This in turn has led to strategic alliances between GCS scientists and business, of which *Eleuthera 2030* is but one recent example.[30]

Projects designed to promote "sustainable solutions" are therefore highly creative, built on alliances with business and based in historical contingency. Their job is to produce consumable scientific fetishes. This is what is at work in the reformation of Eleuthera's "islandness." *Eleuthera 2030* and the investment in island environmental education provided by the Island Academy are forms of pragmatism configured as a business plan and entrepreneurial model for student-tourist visitation and sustainable development. Eleuthera becomes a consumable educational island and performs a kind of translocality.[31] The island is now a model island on the academy's website and the setting of the traveling PowerPoint presentations that demonstrate Island Solutions' responsibilities and innovations in the Anthropocene. Eleuthera is an island that does work for its inventors and investors as it travels in this way.

Further, if the academy's mandate is to inculcate "future leaders" and responsible consumers or investors who understand the environment and can make "good" policy decisions, then who will these leaders be in the Anthropocene era? Anthropologists understand that individuals who promote conservation and the protection of the natural world, what some scholars call "environmental subjects," are the result of specific moments in time, shaped by particular intersections of nature, knowledge, and value. They are highly contingent actors, not just people who have natural traits for leadership.[32] Utilizing this approach allows scholars to consider leadership itself as an effect of powerful socializing practices. Therefore, the events that form transnational environmental movements in the Anthropocene result in changing conceptions of the self and conduct.

Using this rubric, we can see how the Island Academy students' Anthropocenic subjectivity (more than just environmental subjectivity) takes shape. The academy, as a contemporary Caribbean laboratory, educational institution, and research center, explicitly sets out to influence subjectivity with its course work, island-based fieldwork, living systems, and public outreach. As a contemporary revisitation of the Eleutheran Adventurers, the scientists, students, and teachers become the "early explorers" who may open up the Bahamian "frontier" to new forms of scientific visitation, education, settlement, and investment. Over the years I have come to think of the academy as a sort of boot camp for the inculcation of science-based ideas about life and how to live it in the Anthropocene through the generation of visiting adventurers: subjects figured as future leaders, entrepreneurs, consumers, and, significantly, as investors.

Anthropocene subjects are designed to be investors. At the Island Academy students are taught to be leaders whose prowess is based on experiential and self- and place-based knowledge. They demonstrate leadership by sharing knowledge through outreach to various publics and stakeholders. But Douglass, Lamar, and many South Eleutherans easily recognize that part of the expectation of leadership for these students, some of the wealthiest people in the United States, is that they will put their money where their bodies are. Douglass and Lamar hope such people will invest in their sustainable enterprises, while South Eleutherans hope they will invest in property and hotels on their island. Even the Island Academy hopes they will continue to invest in the school and its designs for living, helping to spread the market for this sort of business plan.

The academy students, as Anthropocene subjects, are supposed to use their accumulated wealth, in many cases generated through the extraction of planetary resources, to invest in plans that reframe these same resources as objects for consumption in the name of sustainability. They are taught to value spaces and places that can provide a return on investment as alternatives to enclave resorts or other forms of unsustainable extraction such as sand mining or drilling for liquid natural gas. Their sense of self—their subjectivity—is tied to a sense of responsibility for planetary health as well as the capacity to manage and profit from planetary systems. But this subjectivity is privileged. Just as the visiting students and staff are imagined as *global* venture capitalists and technocratic Anthropocene subjects, so too are island residents imagined as merely *local* island-bound rural subjects through the classes taught at the academy and through preexisting forms of supremacy.

Island people in this story are localized, essentialized, and denied the capacity to design their own sustainable systems. At the academy islanders are construed as "living history" and "people with a past," and it is the U.S. students who have a future, albeit a future that is carefully valued and managed. At the same time some Bahamians participate in that localization by promoting their island and the Island Academy's branding as a field-based education, research, and investment model. They do this for multiple reasons, one of which is that their own hopes for the future of the islands are tied to the speculations of these enterprises.

The resort enclave model and paradise brand proved risky and prone to bust on Eleuthera, but what can rebranding for science-based tourism and sustainable design do to remake the future? Lamar and Douglass are two black Bahamians who came through the Island Academy at different stages of life and who had different experiences there. But they were both inculcated with and inspired by the island business-plan model. On Eleuthera they lived a carefully designed Anthropocene experience, molded within the crucible of the educational island spatial product, becoming advocates for island ecology and global change science, but, unlike their U.S. counterparts, they are not settlers. They are attempting to navigate the slippery line between being localized as living-history lessons for visiting wealthy white U.S. students and becoming young Bahamian environmental entrepreneurs, capable of capturing that capital for themselves.

The educational islands are a kind of travel market for the production of Anthropocene spatial products and subjectivities. Through the operations of the academy, Eleuthera has become a testing ground, neocolonial model, and educational arena for the development of future leaders and the invention of sustainable solutions. Islands like Eleuthera are increasingly becoming branded places where ecobiopolitical designs for the Anthropocene will be realized. Yet there is the other aspect of the education islands. Sustainable designs—those that carry the hopes of an island, a nation, or even a planet—are still embroiled in creating a transformative settlement, in making fields and lives into laboratories, in inventing new practices of place that can still, after all this time, produce ways of knowing that reinscribe difference between visitor and visited, island researcher and island resident, foreign investor and local entrepreneur. The Anthropocene subjects produced in the educational islands do not yet absolve science tourists or international research stations from their connection to the colonial past because they do not yet prevent that past from perpetuating into the future.

3

Sea of Green

DREAMS AND REALITY

Imagine a world where you can't tell where dreams begin and reality ends. This is The Bahamas. And it's like no other place on Earth.

—The Islands of The Bahamas

The islands of The Bahamas are a material reality in that they are an archipelago in the Atlantic Ocean inhabited by about 380,000 people primarily descended from former African slaves, British colonists to the Americas, and Caribbean regional migrants.[1] They are simultaneously an abstract imaginary in that they are a destination brand for a certain kind of dream vacation. The history of Bahamian tourism is a history of branding, marketing, and selling dreams. Dream weaving is, in a sense, a national pastime. The nation's tourism "pioneers" are some of the most recognizable names in Bahamian history and some of the most powerful figures in government. Their descendants are some of the wealthiest members of Bahamian society. A textbook example of this tourism discourse can be found in the book, *History of Tourism in The Bahamas,* and it is useful to spend a little time with that narrative in which the author, a Ministry of Tourism official, describes the country as owing its very existence to tourism.[2]

"Honest tourism" reigns in The Bahamas of the present day, according to the national narrative, and it is often reinscribed into past events of national significance.[3] The long history of arrival and migration in The Bahamas has been appropriated by the Ministry of Tourism, whose staff makes efforts to center Bahamian tourism in the history of the Americas. Christopher Columbus becomes the "first recorded tourist to the Bahama Islands" in the fifteenth century, landing his Spanish ships on the Island of San Salvador, supposedly welcomed by the indigenous people, described as the first "hosts."[4] Beginning with this arrival, important events in Bahamian colonial history are reimagined as precursors to contemporary tourism. Pirates were the "notorious" visitors of the seventeenth and eighteenth centuries; U.S. "rebels" traveled to the Bahamian colony during the blockade-running era of the U.S. Civil War; the British Loyalists in the early nineteenth century set up infrastructure that would later provide the foundation for mass visitation; "revelers" traveled to The Bahamas during the U.S. Prohibition era to drink and mingle in bars, spurring the development of the hotel infrastructure in Nassau; the Duke and Duchess of Windsor put The Bahamas "on the word map" in the 1940s when they arrived to govern the colony, attracting the sophisticated and wealthy international set to the country. The thriving slave trade is notably omitted from this revised history, as are the Duke and Duchess of Windsor's Nazi ties and the Bahamian tendency to mirror Jim Crow–era patterns of U.S. segregation in the hotel industry. It is within sanitized narratives like these that the ministry explicitly reorients the history of the islands around idealized economies of visitation.

Within the official discourse of tourism, the twentieth century is characterized by the development of the industry with a linear teleology culminating in the present era of massive tourism success. In the late 1930s the annual visitor population exceeded that of the native population for the first time. As the "Riviera of the Western Hemisphere," The Bahamas created the Development Board in the 1940s—the precursor to the Ministry of Tourism—charged with the duty of selling the colony. In the 1950s advances in aviation meant more affordable travel and the heavy promotion of The Bahamas as a destination for Americans seeking sun year round. The promotional strategy, developed by Stafford Sands, of marketing The Bahamas as a (racially segregated) beach holiday and vacation paradise, is still prevalent today. In the 1960s there were huge increases in the tourism budget, and the country was known to have a "competitive advantage as an archipelago" in that each island could be visited separately.[5] In the 1970s, after gaining independence from Britain, The Bahamas was the only Caribbean member of the International Union of Official Travel Organizations, leading to the use of statistical tourism measures, global industry standards, and the hire of major research consultancies, such as the Interpublic Group of Companies, defining The Bahamas as a global destination.[6] Such market research identified an "absence of cultural identity" in The Bahamas, which has been a subject of much debate ever since.[7]

In the 1980s tourism was defined, as it is today, as the world's largest industry, and the Caribbean Tourism Organization formed to promote the region to the world. This was the era in which the slogan "It's better in The Bahamas" became widely promoted on T-shirts to the extent that they are still found in U.S. thrift stores, decades later and thousands of miles away. This was also the era in which, to differentiate itself from other Caribbean countries and highlight its "culture," the ministry developed the feature of Bahamian Junkanoo, marketed as a festival of indigenous music and dance, as an example of "culture fully integrated into tourism promotions."[8] In the present, with annual visitor numbers passing six million, The Bahamas tourism industry refers to itself as a leader in continual "product improvement."

In a recent round of product development, The Bahamas' Ministry of Tourism created a new logo for the latest national tourism brand campaign. I cannot present it here because it can be legally displayed only for promotional purposes, but the national logo can be readily found on the internet in Bahamian tourism advertising. The logo appears abstract, evocative of tropical petals or bright droplets of water, but it was cleverly designed by a European consulting firm to represent the islands of the archipelago. The ministry has gone to some lengths to make this association apparent, presenting the logo in conjunction with an interactive map of the islands that switches interchangeably between the logo and map on the ministry's national tourism website.[9] Clicking on the largest yellow "bloblet" brings up an image of a private seaplane parked on a small, sandy beach lined with palms and decked with a single pair of beach lounge chairs. The copy reads, "Andros: Ecotourism Sanctuary. The perfect destination for the eco-traveler, and home to the world's largest fringing barrier reef. Escape reality and enjoy the natural beauty of Andros." A light blue pair of bloblets brings up an image of an empty, windswept beach and reads, "Acklins and Crooked Island: An Escape from Civilization. Remote and the definition of seclusion. Miles of undisturbed beaches, countless coral gardens, and a 1,000 square mile lagoon. What more can you ask for?"[10]

The logo and website are attempts to brand the country as a multisited destination, each island offering a unique experience to the discerning traveler. This is also an attempt to spread the economic benefits of tourism around the country beyond the trap of New Providence—the island with the international airport, major cruise-ship terminal, megaresorts, and the majority of the population. When you click on the central red bloblet on the website, it reads, "Nassau and Paradise Island: Metropolitan glamour, tropical ease. A tropical metropolis, Nassau, the capital city is always buzzing with nightlife, festivals, and excitement. Get loose, break-free, have fun. No one is watching." It is hoped that if tourism can take hold in the Family Islands (known in colonial times as the Out Islands), then people will move back home from overpopulated Nassau and return to their islands of origin to spread the national industry.

Despite the desire to spread the tourism economy around the archipelago, the ministry advertises the depopulated aspect of the Family Islands as a green destination amenity. In the discourse of the ministry, the country has been blessed with a "natural environment that makes The Bahamas one of the best places in the world to live and play," and the "geography and climate have been conducive to the development of an enviable service economy with tourism as its centerpiece." As part of the differentiation strategy, marketing the "natural environment" of The Bahamas is essential, in large part because there are many more possible destinations in the world than ever before, and the competition for visitors has increased in intensity within the Caribbean region. Tourism officials and their partners in the private sector "realized that natural attributes provided a basis for development, and that they could be used in the skillful marketing of the tourism product."[11]

A recent ad campaign on social media sponsored by the Bahamas Out Island Promotion Board exemplifies this skillful marketing. Again, this material could not be legally reprinted here because this book is not promoting The Bahamas as a tourism destination, but the image circulated on Facebook shows a lone woman sitting on a tiny white-sand cove beach nestled between grassy sand dunes next to a startlingly blue ocean under a calm blue sky. The header reads, "84% of The Bahamas 100% unspoiled." The header and the image imply that the Family Islands are tranquil, ecologically pristine, undeveloped, and an alternative to the flashy casinos and rampant commerce of Nassau. This advertisement attempts to reach a sophisticated visitor who will pay a little more to travel a little farther to experience a different kind of Bahamian island. In this marketing vision the Family Islands represent a dream of ecologically mediated tranquility.

Differentiating the islands of the archipelago from one another while holding them all together as one travel brand is a prominent feature of Bahamian marketing. But another theme has emerged in recent years:

small islands in the face of anthropogenic climate change. For the average hedonistic visitor to The Bahamas, the effects of a warming planet are subtle and often overlooked. But global change science (GCS) scientists find that coral reefs are more susceptible to bleaching events, ranging from a lightening of the organisms in their extremities to a complete transformation into shockingly white undersea formations. Conversations with architects reveal that new coastal developments have to raise their foundations by several feet to adapt to predicted rates of sea-level rise and storm surge. And followers of national news know that the Bahamian government has solicited alternative energy plans from international companies, exploring the possibility of solar, wind, or ocean power generation to partially substitute for imported fossil fuels. The emergent effects of anthropogenic climate change are felt everywhere in the archipelago, but they are still on the fringes of perception for most of the visiting population.

One manifestation of the articulation of climate change in The Bahamas has been the nation's participation as one of the Small Island Developing States (SIDS) at the United Nations, a group that has operated since the 1994 Barbados Program of Action.[12] Aligning as one of the SIDS or as a member of the Alliance of Small Islands States allows The Bahamas to enter an international dialogue about the vulnerabilities and risks shared by small islands in the face of anthropogenic change, to have a seat at the table in the negotiations to mitigate this change, and to share any benefits such membership might bring. One benefit is participation in the extensive international education and outreach machinery that coordinates annual workshops in member nations throughout the region through a network of international nongovernmental organizations. The dream-weaving quality of the Caribbean island tourism industry and the stark manifestations of anthropogenic change for small-island states converge in meetings on the intersection of tourism and climate change. These meetings turn The Bahamas, and small-island regions like it, into Anthropocene Islands floating in a sea of green, where the environment and commerce meet.

BRANDING OPPORTUNITIES

In 2008 I attended a regional workshop on climate change and Caribbean tourism, where it wasn't clear where dreams began and reality ended. The event was hosted by The Bahamas' Ministry of Tourism, the Caribbean Tourism Organization, and the Caribbean Forum of Afri-

FIGURE 8. The Wyndham Hotel resort ballroom, Nassau. Photo by author (2008).

can, Caribbean and Pacific States' Caribbean Regional Sustainable Tourism Development Programme. Attendees consisted of representatives from regional tourism and hotel organizations, private business, Bahamian environmental governmental and nongovernmental organizations, members of Parliament, and members of the press. Arriving at the hotel convention center in Nassau, I was impressed by the number of attendees and the fact that dozens of men and women had traveled from all over the region to be there.[13] I was also impressed that there was a representative from every Bahamian environmental nongovernmental organization in attendance. The setting was grand, with uniformed hotel staff clearing away used dishes and refreshing drinks. I understood that this was a relatively well-funded event compared to most environmental meetings in the country, and everyone appeared to be taking it very seriously.

I was immediately struck by the relationships among attendees that clearly preexisted the meeting. Nearly everyone in attendance had been formally invited, and they were quite familiar with periodic events such as this.[14] What's more, the attendees were also quite familiar with one another. As I hovered awkwardly by the coffee and pastry station during the opening meet and greet, I became cognizant that officers of the Bahamian chapter of the Nature Conservancy were well known to members of the Jamaican Ministry of Tourism. The head of a local nongovernmental organization was friendly with a representative from a regional research station. People were hailing one another warmly all around the glinting ballroom. These actors who are repeatedly at odds over the scope and degree of environmental conservation and economic development in the region are actually all part of a shared community concerned with the region's future, and they have been assisted by international institutions in working together on regional policy for some time.

Soon we were asked to take our seats, and the light from the glittering chandeliers dimmed. The "workshop" turned out to be a series of talks on the effects of climate change in the region and the subsequent dangers to the tourism industry that is so dependent on socially and ecologically stable Caribbean islands. The one talk that still stands out to me was titled "Climate Changing the Industry? The International Policy and Market Response to Global Warming and the Challenges and Opportunities That Climate Change Issues Present for the Caribbean Tourism Sector." The Caribbean Tourism Organization had invited a British academic consultant, unknown to the audience, to deliver a presentation on climate change adaptation and mitigation in the tourism industry. In his lengthy talk, delivered with strong conviction, the consultant made a number of connections, linking the science of the Intergovernmental Panel on Climate Change (IPCC) with future plans for the Caribbean and the Bahamian "tourism product."

The consultant's talk relied heavily on GCS research. It included an overview of current climate change science, showing charts and graphs published by the IPCC, which described the increase in global atmospheric temperatures and the projected increase in surface temperature over the course of the twenty-first century. This research and the scientific projections were incontrovertible, he asserted, but so was the message for the region's tourism industry.

In a segment about new realities for tourism in a world affected by global change, the consultant identified the Caribbean region as a "tourism vulnerability hotspot," characterized by IPCC-predicted warmer summers, sea-level rise, increased extreme weather events, freshwater scarcity issues, marine biodiversity loss, an increase in disease outbreaks, political destabilization, and travel cost increases. Caribbean tourism was described as "crucially interdependent" on climate, forming the basis for the economic development and local livelihoods in most Caribbean nations. Sea-level rise, the increased rate and strength of Atlantic hurricanes, coral bleaching, coastal erosion, and changes in rainfall patterns were described as detractors from the "natural resources of tourism," all of which would lead to the "loss of attractiveness of the region as a destination, the loss of employment in the industry, increased industry insurance costs, increased operating costs, and changes in patterns of tourist travel flows." Citing vulnerability rankings for island and coastal states, he identified The Bahamas as one of the region's most vulnerable countries, highly susceptible to this potential loss of attractiveness to the tourism industry.

The consultant departed from those who had spoken before when he described the world tourism sector's contributions to global greenhouse gas emissions, 75 percent of which stem from the transportation of tourists around the world, calculated to be 5 percent of all world emissions at that time. This placed the world tourism industry as the fifth-largest emitter after the United States, China, the European Union, and Russia.[15] Tourism was construed as both a "vector and a victim" of climate change because, although the Caribbean has not significantly contributed to global warming through the production of industrial carbon dioxide emissions, the region's tourism industry most certainly has. The economies of the Caribbean were described as complicit in producing high levels of travel-related emissions.

The take-home message of the presentation became clear when the consultant arrived at his recommendation: the "carbon neutral destination" as an adaptation and mitigation strategy for small-island destinations. For a location to continue to sell itself in a tourist market that has been constrained by carbon, tourism officials must perform three steps: (1) they must measure and assess emissions produced by visitors' travel, accommodations, and activities and devise means to reduce them within the country; (2) they must "decarbonate" by switching to energy-efficient practices or alternative energy sources; and (3) they must offset their remaining emissions using offset purchasing options approved by Gold Standard Certified Emission Reductions registered with the United Nations Framework Convention on Climate Change. Tourists and other travelers, the consultant assured us, would be willing to pay a small offsetting fee as part of their travel expenditures, knowing that they were contributing to the state of island carbon neutrality that they were visiting. The consultant noted the great "Caribbean potential" for rebranding resulting from the fact that this is a "unique and ground-breaking time for tourism," and he touted the development of "sustainable destinations" as crucial for maintaining economic development in the region.

The consultant clearly assumed that becoming a carbon-neutral destination might be an attractive product for a tourism-dependent small-island country like The Bahamas. He sold the idea as having the potential for "greening the image of the country and the region." Institutions in The Bahamas are already quite savvy when it comes to branding the country as a particular kind of destination, but the British consultant presented the Ministry of Tourism with a novel opportunity to diversify the tourism product away from the "sun, sand, and sea" brand to a marketing strategy that sells the very vulnerability of the islands themselves. In this strategy explaining how visitation negatively affects the country's fragile resources and asking visitors to pay a fee to protect these resources may actually attract more visitors on a carbon free vacation."[16] In this case the threat of climate change is not just a research opportunity for the GCS community. It is also a means to make the region a leading example of visitable approaches to climate change mitigation and adaptation measures. All it would take is for Caribbean countries such as The Bahamas to rebrand themselves in terms of offsets, alternative energy, and carbon neutrality.

SMALL ISLANDS, CLIMATE CHANGE, AND (SOCIAL) SCIENCE

The British consultant's remarks relied on an existing body of GCS work that, like the tourism industry, frames small islands as distinctive spaces. Small islands have become a powerful general category for international scientists in the context of climate change, and GCS participates in producing the small-island state as a fragile political object based on expanding notions of island vulnerability. The institutional discourse of science, policy, and tourism all work together to validate the international planning protocols that link small-island states and climate change as conjoined phenomena. It is therefore especially important to understand that the science of climate change is itself a driver of culture and society within the crucible of the Anthropocene idea.

As the workshop demonstrates, when it comes to climate change and small islands, the discourses of crisis, vulnerability, sustainability, and development opportunity coexist. This convergence is due in part to the fact that climate change impacts have been framed as inevitable but also, importantly, as predictable by the IPCC, the highest GCS authority on the issue.[17] This predictability allows for the proliferation of preparedness protocols and for the linkage of climate change with development concerns, such as tourism. Sustainable development for small islands is redefined in the context of climate change in terms of adaptation to climate impacts on ecosystems, food production, and economic growth. Small-island states are now construed as less likely to meet sustainable development goals unless they incorporate climate change scenarios and adaptation measures into their national development plans. These evolving circumstances beg the question, what does it actually mean to refer to a place as a "typical Small Island Developing State"?[18]

In general, it is likely that the small island owes its roots to international economic development forums and the anthropological study of island peoples, rooted in biological notions of evolution and isolation. In fact, small islands are amazingly diverse in terms of population size, economic wealth, culture, and language.[19] While there is no single, agreed-on global list of small islands or small-island states and no single system of classification, there are currently forty-four member states and observers of the Alliance of Small Islands States.[20] Additionally, the SIDS Network, a political category created in Barbados in 1994 at the first Global Conference on the Sustainable Development of SIDS, has currently classified fifty-seven states.[21] A common general characterization of small islands and small-island states is that they have a small land area, a small population, limited natural resources, undiversified economies, and weak institutional capacities. They must contend with spatial remoteness, vulnerability to economic shock and natural disaster, and fragile ecosystems.[22]

Small islands are further categorically institutionalized in the Anthropocene in specific ways. GCS typically views climate change as a global event that threatens the economic and social stability of small-island states and demands international attention and research.[23] The predictions are dire: two to three commonwealth nations will disappear by the end of the century because of sea-level rise; the Maldives are under threat of complete submersion in the Indian Ocean; the Caribbean region may lose low-lying coastal plains; some of the islands of The Bahamas are under threat of submersion; and, in the Pacific, small-island nations such as Tuvalu have begun population-migration agreements with larger nations such as New Zealand.[24] Furthermore, while the loss of territory is not predicted for all small islands, enormous economic costs are seen as the most severe threat to small-island stability. This news is especially egregious when contrasted with the argument that small islands have not directly contributed to the production of greenhouse gases to the extent that larger industrialized nations have. In other words, tourism aside, many low-lying small island regions, governments, and people are set to suffer extensively from a climate situation that they had no substantive part in creating.[25]

This inherent inequity between large industrialized states and small-island or small developing coastal states has inspired block political organizing around climate debates within the United Nations. The Alliance of Small Islands States was formed in the 1990s to provide a platform for small-island concerns, and small islands have now become an ethical orientation in international climate debates. They represent the unjust nature of the climate crisis in the Anthropocene, and identification as a small-island state can be a tool for obtaining political leverage.

How do social scientists, anthropologists and otherwise, follow these developments? What role should social science take within these problems? How can social science open up new relevant conversations? One possible answer is that social scientists can study the way that GCS practices reorient the very meaning of life in small-island places. Social scientists can trace the development of emergent scientific worldviews in the Anthropocene as they coexist with or modify existing worldviews. They can also examine GCS ideas, like climate change impacts and vulnerability, that produce small-island states as distinct locations with opportunities for related enterprise.

There is a loose body of social research on small islands that is embedded in a "vulnerability paradigm," describing island social realities as existing in a state of decline from rapid population growth and the spread of inefficient economic development schemes.[26] Another trend involves social scientists advocating for the incorporation of local people into small-island scientific research, modeling, and development planning in the form of human behavioral variables.[27] This social science advocates for the revision of international methods of assessment and management to "better fit" small-island state characteristics.[28] Made within governance circles, this work is oriented toward making better policy decisions.

Despite these trends, one of the advantages of anthropology is its vantage point from outside the management policy frame and its ability to produce knowledge that does not fit a priori into institutional molds, such as the small-island state and its supposed adaptation and management needs.[29] The strength of anthropology still lies in long-term fieldwork and embedded relationships with both research subjects and research questions. Anthropology can take the issue of climate change beyond the terms of adaptation, vulnerability, and resilience to challenge the practices of scientific research itself.[30]

The small-island developing state is thus a charged political category that takes shape, in large part, within the policy frameworks of GCS predictions and policy. Yet there is more for social science to do than be only "policy appropriate," and social scientists must examine what the policy-negotiation process itself makes possible and impossible, even in low-level climate and tourism workshops. Unfortunately, the checks-and-balances role of social science is not always seen as relevant to climate change organizations that deal with adaptation, mitigation, valuation, and policy. This is because some critiques cannot ever be made feasible or actionable in policy programmatic ways. The relevance of anthropology lies in its ability to describe other possibilities and to create an alternative space for thought. That kind of work, the kind advocated for here, is a purposefully untimely anthropology of climate change.[31]

SMALL-ISLAND VULNERABILITY

> The vulnerability of The Bahamas to the impacts of climate change is well known given its geographical vulnerabilities (limited land masses, low-relief and dispersion of islands), environmental vulnerabilities (high temperatures, storm surges, sea level rise, flooding, tropical cyclones and non-tropical processes), the concentration of socio-economic activities and critical infrastructure in narrow coastal zones, and its dependence on tourism and the limited human and institutional capacity. The Bahamas is also highly dependent on the imports of fossil fuels for energy needs, which places a heavy burden on its economies as a result of the vagaries of global petroleum prices. Based on the available scientific consensus we can expect more frequent and intense impacts over time. It is within this context that The Bahamas is expected to adapt to the impacts of climate change while at the same time pursue a low carbon pathway in conformity to growing international and public pressure for environmentally friendly development that reduces their "carbon footprint" and exposure to climate change, while also increasing energy security.
>
> —Government of The Bahamas

Vulnerability is a constitutive part of the small-island category. Small islands and vulnerability have long been linked through the discourse of economic development and environmental hazards, and development projects set the precedent for assessing small-island states with economic vulnerability indices.[32] The use of vulnerability indices for planning purposes and prioritization seems to have taken off in the mid-1990s. The institutionalization of climate change within international development policy has done nothing if not increase and refine this "rubric of vulnerability."[33] Indeed, the British consultant at the climate and tourism workshop relied on assessments of Bahamian island vulnerability to the impacts of climate change ("some of the most vulnerable islands in the region") to make his pitch for the promotion of the region as a carbon-neutral destination. But what are the implications of promoting vulnerability as a small-island socioecologic?

The editors of the *Third Assessment Report* of the IPCC define vulnerability as "the degree to which a system is sensitive to, or unable to cope with, adverse effects of climate change, including climate variability and extremes."[34] And the IPCC's *Fourth Assessment Report* states, "Vulnerability to climate change is the degree to which these systems are susceptible to, and unable to cope with, the adverse impacts."[35] Vulnerability is therefore a vague term, but this vagueness allows for a great deal of work and organizing to be done in its name as a keyword in the Anthropocene. Indexing and ranking are two central aspects of the vulnerability paradigm.

The procedure of the UN Framework Convention on Climate Change for dealing with global SIDS planning includes an explanation of vulnerability assessment.[36] Vulnerability is assessed by identifying the severity of short- and long-term climate impacts and by prioritizing physical and economic impacts for concentrated action.[37] Vulnerability assessment was created to generate appropriate indices for SIDS that represented more than just the loss of national GDP. National vulnerability indices are intended to pinpoint the institutional weaknesses of SIDS.[38] The risk of natural hazards and the fragility of economic and ecological systems therefore became key components of vulnerability indices.[39]

Vulnerability indices allow for comparison between entities and set categories, and there are currently multiple vulnerability indices and methods for the assessment of vulnerability. While some argue that small-island vulnerabilities and climate-led disasters matter primarily in terms of their economic implications, others have been attempting to better integrate the analysis of ecology and economy.[40] No one index measures the vulnerability of all SIDS or all states. Some indices measure sea-level rise vulnerability, while others measure economic or environmental vulnerability. There is even a Commonwealth Vulnerability Index and an Environmental Vulnerability Index.

In terms of ranking the vulnerability of small islands—a major concern at the turn of the twenty-first century—the Maldives has been shown by some indices to be the most vulnerable in terms of sea-level rise, while all the small islands in the Pacific are categorized as extremely vulnerable. The Commonwealth Vulnerability Index calculated Vanuatu as the most vulnerable nation in the world, with seventeen of the most vulnerable twenty-five being small-island states. Small islands in the tropics are considered to be especially vulnerable to warming temperatures, despite the fact that the most extreme temperature rises are occurring at the poles, and atoll islands are placed near the top of most indices.[41]

In the Caribbean the impacts of climate change are identified as erosion from sea-level rise, high-water tables, property loss, threats to tourism infrastructure, and flooding of low-lying areas.[42] Other effects are the warming of coral reef habitats, shifts in rainfall affecting agriculture, landslides, and increased hurricane strength and frequency. These "impacts" matter because the Caribbean region's systemic vulnerabilities have been identified, categorized, ranked, and indexed to develop national targeted corrective action plans. Key vulnerability issues for the Caribbean region include, among others, the intensification of water scarcity, unstable human settlement and infrastructure, threats to the viability of the tourism economy, increased fragility of coastal zone and coral ecosystems, increasingly fragile fisheries and agricultural sectors, the growing prevalence of tropical diseases, and exposure to storms and hurricanes.[43] The Caribbean region has also been identified as the most economically volatile region in terms of consumption, as vulnerability is linked to public debt.[44] The broad calculation appears to be simple: climate impact plus high island vulnerability equals state disaster.

One result of these indexing exercises has been the development of the idea that the main commonalities of small islands consist of the possibility of disaster, the existence of multiple vulnerabilities, and the sense that small islands have a low capacity for resilience in a number of sectors. Vulnerability has become recognized as an inherent condition of small islands: it is said to preexist the threat of natural hazards, making small islands susceptible to all hazards. Indexing also shows that vulnerability has been grouped into social, economic, and environmental categories.[45] As a result, small islands are construed as naturally vulnerable, structurally vulnerable, and systemically vulnerable.

Much like regional tourist branding, part of the very purpose of vulnerability assessment is to influence public perception.[46] Vulnerability indices were developed along with the awareness of the climate crisis to make explicitly visible the disadvantages of small islands and small-island regions. This form of policy programmatic assessment has now been repeatedly strengthened with each international meeting on the topic of small islands and climate change. The main benefits are seen to be attention to the issue, easy comprehension for policy makers, and guidance for the allocation of scarce funds.[47] These assessments promote an understanding of vulnerability that is supposedly politically expedient.

I would like to introduce a healthy skepticism with regard to the popular use of the vulnerability paradigm as a trope in the Anthropocene. We need further inquiry into the implications of reifying the so-called inherent vulnerability of small-island places in terms that appeal to policy makers. Although climate change concerns have expanded the role of vulnerability assessments, they are still primarily economic in focus and based on narrow quantitative analyses. Vulnerability, in the global context of climate change, has also linked small islands to generalized and homogenous ideas about economic, environmental, and social systems that lack historical specificity. In addition, there are now "expert vulnerability practitioners" whose job it is to calculate, rank, and prioritize island vulnerabilities. There has yet been little research done with these populations of emergent experts, especially as they pertain to the codification and classification of the category of small-island states.

Some anthropologists have recently called for rethinking the "rubric of vulnerability," which is top-down, denying local agency and an awareness of the meaning making that proliferates around climate change and island issues.[48] They ask us to recognize that climate change affects different people differently and that vulnerability is tied to histories of domination and social inequities that are not at all readily apparent in most vulnerability indices.[49] Another danger is that small-island communities become reified as lacking capacity or power.[50] Vulnerability rankings can alter the decisions of aid donors and investors, and the threat of climate change impacts might do more harm in some cases than the actually experienced impacts themselves.[51] What's more, vulnerability indices presuppose what is valuable for different groups. Indices assume that inhabitants of islands share vulnerabilities in equal measure, when in fact there may be infinite variations to susceptibility within islands and between social groups and individuals. Vulnerability assessment is potentially profoundly exclusionary.[52]

Social observers must interrogate the way islands are reimagined in the Anthropocene. We must examine the mainstreaming of vulnerability to consider who benefits from these calculations and identify what is made possible or impossible within this socioecological paradigm. Vulnerability indexing is a GCS tool that contributes to the production of Anthropocene spatial categories and Anthropocene products. In the context of anthropogenic climate change, such observations are complementary to efforts to make ethical social decisions, and they can allow alternative and silenced realities to enter into the frame of debate.[53] This sort of untimely observation can provide a corrective to policies that reinforce dangerous development notions of island insufficiency, weakness, homogeneity, dependency, and isolation.

SIDS IN THE SEA OF GREEN

The "small island" is a category sparked by the climate crisis and the institutionalization of the vulnerability paradigm. And small islands are now concrete manifestations of Anthropocene socioecologics. The Anthropocene is the context in which small islands are able to lay claim to—or request adaptation and mitigation aid from—the international climate community, specifically organizations concerned with issues of climate justice between the developed and developing world. Small islands also reveal how Anthropocene crises become unexpected opportunities for some actors. The British consultant exemplifies the drive to link climate change adaptation and mitigation to the expansion of travel markets as a sustainable development tool. Tourism is therefore a good position from which to observe this mixing of GCS science and the generation of capital. For those interested in contemporary small-island issues, cases like these exemplify the problems of the day.

From this perspective we can see that international workshops, like the one I attended in 2008, help reproduce the Anthropocene Islands through active experimentation with dreams. In this case The Bahamas has become entangled in an Anthropocene assemblage of international institutions, climate models, GCS scientists, bureaucrats, consultancies, and the rubric of vulnerability.[54] The Bahamian "tourism product" is part of this entanglement. Climate could very well become a more and more explicit part of a small-island marketing package designed around visiting one of the most vulnerable countries in the Caribbean region. Although it is a government body, the Ministry of Tourism is run as a business, and this business has the potential to solidify such possible local realities.[55]

Within the marketing dreams that link the Bahamian archipelago, climate change, and the Ministry of Tourism, local realities and histories have been reimagined.[56] The Bahamas is already perceived in the form of a carefully designed multidestination tourism product, historically branded as paradise in the past and present, but remaking the

Bahama Islands into a climate-conscious Anthropocene product implies rebranding the paradise as vulnerable. This means that collaborations between climate science and the tourism industry are now a part of the process of orchestrated distinction that creates locality and animates markets, brands, categories, regions, and islands.

Distinction is necessarily the quality of being in a relationship.[57] Just as each island of The Bahamas as a potential local tourism product is always in a relationship with The Bahamas as a national product, The Bahamas, as a national tourism product, is always in a relationship with the Caribbean as a regional tourism product, with the Ministry of Tourism constantly engineering distinction from it, positioning the country and its islands in a social system of difference.[58] The creation of the small island as a political category, the listing of islands in the world that belong to that category, and the ranking of those islands and regions in terms of anthropogenic vulnerability mirrors these existing brand patterns that attempt to strategically produce distinction in The Bahamas and the Caribbean.

The brand of sun, sand, and sea is one example of the engineering of distinction that has lost its edge for the ministry, as the brand now represents the entire region and is no longer the exclusive domain of The Bahamas, if it ever was. Yet ministry officials recognize that marketing The Bahamas through "honest tourism" brings attention to the region and vice versa, because branding the Caribbean region also influences the collection of images and emotions associated with The Bahamas. Similarly, science-based climate-conscious branding of The Bahamas as one of the world's vulnerable small-island regions and then subsequently as the world's first small-island carbon-neutral region—as a veritable sea of green—might influence these dynamic systems of difference in ways that cannot yet be known.

What is important here is neither the conviction of the foreign consultant nor the extent to which Caribbean governments will eventually incorporate his recommendations to mitigate, decarbonate, and offset into their national policy frameworks. In fact, The Bahamas is far from committed to a carbon-free future, and it is readily apparent that the national industry has a large carbon footprint. Bahamians and tourists lead carbon-intensive lives, even if emissions in The Bahamas are dwarfed by most industrialized nations and nations with larger populations. The national electricity grid that powers all the homes and businesses and all the lights and air conditioning at all the hotels runs on crude oil delivered in tankers to a crumbling pier with a postapocalyptic aesthetic. This pier leaks oil into the surrounding sea, which then has to be treated with chemical dispersants. Members of the New Providence conservation community tell me they are not sure which is worse for the marine ecology and human bodies—the oil or the dispersant. The planes and cruise ships that bring the millions of arrivals each year all run on large amounts of fossil fuels. There is no shortage of imported motor vehicles in the archipelago, and most run on expensive gasoline, including the tour vans and jitneys. Driving down Bay Street in Nassau means rolling up your car windows if you are unlucky enough to find yourself behind a cargo truck belching nasty diesel fumes. Besides the sailboats, most other vessels, from party boats to interisland ferryboats, run on gasoline or diesel fuel. Any real transformation toward a carbon-free energy infrastructure remains a long way off and, in any event, would depend on far more than a two-day workshop.

Instead, the consultant shows that a potential Anthropocene product—the dream carbon-neutral island destination—has grown out of GCS research to the extent that it could feasibly be presented at a workshop with an audience of industry professionals, government bureaucrats, and nongovernmental organization employees. The consultant at the workshop demonstrates the economic development opportunities involved, as well as the creation of new forms of market expertise, new research agendas, and new regional distinctions based on the inevitability of global, regional, and local planetary change. Examining these midlevel discussions illustrates the ways climate science legitimates potential business opportunities and allows for the creation of new categories and experts in the Anthropocene. If small-island regions are always dependent on some degree of island product distinction, then science-inspired marketing strategies have the potential to carry real weight in the tourism industry.

I would like to reiterate the question asked earlier in this chapter: what does it mean today to refer to any place as a "typical SIDS"? What are the implications of this categorization? What are we perpetuating when we use this language, consciously or otherwise? How do political categories become aligned with brand platforms? How does crisis also become a platform for the creation of new products? Thinking about small islands and anthropogenic climate change requires that we question the creation of crises and categories like these.

Within the dreamlike vision of the carbon-neutral Caribbean tourism region presented by the visiting consultant, the possibilities almost seem to outweigh the potential threats of climate change. Some years later many of the Bahamian attendees of that 2008 workshop continue to agree that these opportunities can come only in the form of economic expansion under the aegis of tourism. While carbon neutrality is not yet an amenity that distinguishes The Bahamas or the region in the international tourism market, in 2016 the notion that the multidestination tourism product of The Bahamas is the best means to justify ecological protection was alive and well.

In the spring of 2016, I attended the Bahamas Natural History Conference, held at the University of The Bahamas in Nassau. As I listened to panels on topics ranging from forestry to fisheries and from marine mammals to ornithology, I was struck by how often foreign natural scientists would stress how their local research on vulnerable species threatened by anthropogenic coastal development would benefit the tourism industry of The Bahamas at their particular research site. Creating protected areas containing endangered piping plover or Kirtland's warbler populations could attract tourists to see the rare birds in their winter range. Promoting certain islands of The Bahamas as prime habitat for threatened butterfly and moth species could be good for ecotourism. Unique island ponds that contain the world's highest density of seahorses can be protected from conversion to yacht marinas because their tourism potential could bring in alternative forms of revenue. Through it all I was reminded of a sharp comment made by a Bahamian nongovernmental organization representative to her colleagues at the close of the climate change and tourism workshop eight years prior: "Isn't it a bit ironic that they want to combat the rise in tourism-related greenhouse gas emissions in the Caribbean through the promotion of more tourism to the Caribbean?"

This point speaks to the limits of the continual promotion of international tourism expansion in the region. It is one thing to say that the tourism industry can provide an economy to island populations that are the victims of inherent geographic vulnerabilities to climate change. It is another thing altogether to ignore the fact that Caribbean islanders have been shaped by generations of social and historical inequities in which tourism has played a major polarizing role. I would add another irony here that is now familiar: rebranding Caribbean islands to capitalize on their "inherent vulnerability" to climate change does nothing to address or halt the histories of exploitation that helped create that vulnerability. The Bahamas is vulnerable in many ways, but much of this vulnerability is contingent on exploitative colonial economies and the current dependence on foreign investment in large-scale tourism and infrastructure. This is a development model that enriches a few and employs many at a low level in a trickle-down economy that privileges visitors, consultants, and elites over most island citizens. Pointedly, the Bahamian anthropologist Nicolette Bethel has said, when speaking about recommendations made by foreign consultancies, "What The Bahamas needs are not more experts from the outside who come along with solutions that treat us as though we are cookies cut from some other reality," a reality blind to the fact that these experts come from places that are complicit in the very problems in question.[59]

Small islands and the people in them do not exist in isolation from one another or the rest of the world, but neither do they exist as singular categories.[60] Thinking in categorical terms of small-island states can obscure other relationships between people that have been forged over time and through space. As they stand now, climate change response policies—based primarily on nationally bounded frameworks and assessments, a standard set by the international community in the name of national development, sovereignty, and management—can do little to manage island events. Climate-induced migration, foreign investment, and even climate-conscious travel marketing are influenced by the very framing of island vulnerability as ahistorical and disconnected, represented in indices for easy comprehension. One result of such international classification and indexing is the creation of homogeneity in terms of international and national responses to anthropogenic climate change and the understanding of the stakes and opportunities involved. But islands, even small ones, are not homogeneous.

Future international meetings of the climate community could result in the creation of international legal mechanisms involving climate impacts, the redefinition of international human rights standards in terms of island submersion, the loss of territory or state viability, and the designation of environmental refugee status. We may indeed see a renegotiation of the grounds for international asylum and migrant status, a redefinition of the rights of island citizens, a redetermination of when a state can be declared nonexistent, a recalculation of the life expectancy of states, the reclassification of climate change as a potential state of public emergency, the incorporation of environmental law into human rights law, and an expanded definition of adaptation to include more radical options.[61] These possibilities might inspire hope, and these conversations are happening in the Pacific region, but they are as yet largely absent from the rubric of vulnerability in The Bahamas and the perpetual focus on the stabilization of its tourism product.

In The Bahamas the conjunction of climate change and tourism will likely perpetuate the status quo in terms of social inequalities engendered by the tourism industry. But at the same time new institutional roles may be created to accommodate climate-centered business plans. The codification of vulnerabilities for small-island states and the creation of frameworks for national planning have become ends in themselves in the Anthropocene. These are powerfully creative tools. Global change, including climate change, is nothing if not ecobiopolitical, from the realm of international negotiation to the way people conceive of their relationships with others. There are conceptual innovations running all through these discussions, especially in terms of how institutions think about appropriate state actions and categories.

We need more research on the use of vulnerability indices and the classification of small-island states in the name of climate change. This work can document dreams for redevelopment under the auspices of the Anthropocene idea. Such work can open up climate change–based development policy and its experts, practitioners, and audiences for examination, as well as the tools that they produce. We must describe those whose profession it is to observe the world and to remake the world through observation, that is, to observe contemporary society through contemporary society.[62] Anthropologists can examine rebranding campaigns that market small islands as vulnerable Anthropocene spaces. Such scholars have the ability to recognize the explicit targeting of populations and the reimagination of specific locales, as well as the revision of history in the name of development, adaptation, mitigation, vulnerability, and resilience.

Scientific, political, and touristic realities matter in small-island settings to the extent that they can be said to make small islands themselves. Social sciences like critical anthropology can provide the necessary attention to the symbolic and material practices that define mainstream events, an effort that is not currently reflected in many popular climate change and small-island debates. Natural scientists and tourism officials could be more aware of their regional and island-making practices, and international organizations could be more careful regarding the generative power of their categorical language and the tepid utility of their proposed solutions. It is not possible to "offset" a historically polarized and highly unequal economy, after all. Yet dreams for new Anthropocene products and brands, forged in the relationship between small islands, the tourism industry, and GCS-defined climate change—that sea of green—continue to reshape our contemporary world.

4

Aquatic Invaders in the Anthropocene

A FISH STORY

There are rare times during long fieldwork visits when you don't have a clear idea of what you should be doing on a given day.[1] You have no interviews lined up, there are no informant projects to observe, and you are caught up on your note taking. I attempt to solve this dilemma by walking, hoping for some peripatetic inspiration or for some kind of meaningful encounter. In the summer of 2014, on one such walking afternoon in southern Abaco, I found myself at the end of an empty wooden dock in a harbor marina. If you aren't getting on a boat there isn't much to do at the end of a dock except look over the side. I thought I might glimpse a sea turtle or a shark in the clear water, but I was startled to see a fish. It isn't unusual to see fish swimming around docks in The Bahamas, but this was not the usual snapper commonly found schooling in marinas, darting for scraps falling in the water during fish cleaning. This was a strange, boldly striped fish. And it was frozen, hovering upside down. Was it dead? On closer inspection I could see its elaborate fins slowly waving in the water, and I realized that this was a fish I knew well. It was a legendary lionfish, hanging motionless, waiting for prey. I looked for more and easily found them clustered around the pilings in disconcerting groups of three or four. These watchful fish become an all-too-familiar sight on my walks that summer.

A focus on lionfish reveals several things. These creatures reveal how the Anthropocene idea functions in fields beyond climate change. Lionfish

FIGURE 9. A cooler of dead lionfish caught in a tournament, Nassau. Photo by author (2014).

science rearticulates an earlier research focus on fisheries regulation and common pool-resource management centered around fishing communities, complicating the imagined relationship between those whose livelihoods stem from the sea and those whose lives are harvested from the water: the relationship of fisher and fished. Lionfish also reveal the ways global change science (GCS) marine management frames relationships among people, natural resources, and animals within fisheries in the Anthropocene. Within GCS narratives the lionfish fishery is conceptualized as a socioecological relationship between humans and nonhumans that naturalizes certain formations of fish and fishers to create a specific "guilt-free" product to the detriment of a more complex understanding of fisheries. The lionfish is thus an iconic species for the Anthropocene.

I first heard of this strange fish in 2008, when I assisted a young Bahamian graduate student in her field study of lionfish attraction to artificial reefs around the island of New Providence. The study was the main component of her thesis for a master's degree in zoology at a Canadian university, and I helped the student find her reefs in the shallow water, allowing her to save energy as she swam in heavy dive gear behind our boat. From my vantage point on the bow of the twenty-foot Bahamian Department of Marine Resources (DMR) vessel, I could make out the

FIGURE 10. Lionfish swimming upside down in a marina, Abaco. Photo by author (2014).

distinctive blob of an artificial reef on the sandy bottom and point it out to her in the water. The reefs, made of cinder blocks bundled together, were laid out in a grid pattern on the sea floor far enough apart that they could not be seen through a mask in the clear water; it was startlingly easy to swim past one. The student hoped that the maroon-striped and fantastically finned lionfish might be found lurking in these reefs, which would help prove her hypothesis that lionfish are attracted to artificial structures, such as wreckage and boat debris, in near-shore waters. At each reef she would dive down and make note of what was found there: each species and their size, numbers, and density.

The lionfish is beautiful and enigmatic as well as dangerous. Long, wavering spines protrude from the airy fins on its back and belly, each spine loaded with venom. A sting from a venomous spine is extremely painful and potentially life threatening for the very old, the very young, or the sick. Some Bahamian fishers say it will put you on your back for a day. But poison is not what makes the lionfish such a threat or the object of scientific interest. The lionfish is unwelcome in Bahamian waters because it is an invasive marine species.

Indo-Pacific lionfish moved from the category of introduced species to the rare category of "marine fish invasive" in The Bahamas when it was discovered that they reproduce rapidly, eat voraciously, compete for reef space with native fish, and have no natural predators in Bahamian waters.[2] First observed in The Bahamas in the mid-1990s, lionfish established themselves in no time and soon became a common sight for fishers and recreational divers, rapidly increasingly since 2006.[3] These observers were wary of the venomous fish, but when rumors spread that someone had found a dead Nassau grouper in the belly of a lionfish, it quickly became a threatening figure, contributing to the ruin of an already vulnerable fishery and a coral reef environment that some scientists fear has become increasingly "invasible"—that is, susceptible to invasion as a result of human activities.[4] The student's artificial reef experiment was designed to test this notion of environmental invasibility—to discover just how vulnerable the anthropogenically polluted and debris-strewn near-shore waters of the central Bahamas might be to this curious fish.

Thinking about lionfish in The Bahamas as anthropogenic invaders in a vulnerable archipelagic fishery has a remarkable parallel in descriptions of regional overfishing, descriptions in which fishers become the frightening invasive species imperiling the Caribbean's collection of rapidly disappearing ocean life.[5] This increasingly vulnerable fishery has become a scientific object, threatened by fishers and invasive fish who are said to magnify the vulnerability. In this framing of Anthropocene logic, ecological and social-scientific fieldwork narratives and practices define fisheries, fishers, and fish as entities with specific traits. At the same time marine conservation efforts make the lionfish into a compound creature, at once a recent arrival to the islands, a threat to Bahamian biodiversity and the marine resource economy, and an iconic species for science concerning biological life and social categories.

By 2010 lionfish-cooking demonstrations were popping up around the capital, and lionfish was on the menu at number of local restaurants. I visited the DMR office in Nassau to find out why. The office is located in a short yellow government building across Bay Street from the harbor with a view of the old bayside conch middens and the many small fishing boats haphazardly surrounding Potter's Cay Dock. In the cubicle of a Bahamian fisheries officer, I asked why the DMR was promoting the commercialization and consumption of lionfish. The officer thought for a moment and then put it plainly: since the lionfish is a common food item around the Indian Ocean, why not put it in danger here? If Bahamian fish are already in danger from Bahamian fishers, why not let Bahamian fishers do what they do best in this case?

This fish story shows how Bahamian waters and sea life are localized, universalized, and naturalized to legitimate marine environmental management.[6] Environmental managers like the DMR and marine nongovernmental organizations have become the primary creator of available terms with which to think about the connection between humans and aquatic organisms that constitute contemporary fisheries. But their framings put fishers at a decided disadvantage. The story of the lionfish in The Bahamas can show us such framings work to make the Anthropocene Islands. The story highlights ways that Bahamian fishers have been held responsible for fishery health by some members of the science and management community and how this responsibility has become self-evident in that community as a problematic form of socioecological relations. The story also utilizes key national tropes: visitation and invasion. The Bahamas, like many Caribbean islands, has a conditional vocabulary of belonging.[7] *Citizen, resident, expatriate, tourist, immigrant, guest, illegal*—these terms organize social and political life. Lionfish are linked to these tropes within Bahamian fisheries management, and these linkages play into the debates about anthropogenic impacts that both shape and manipulate stakeholders, managers, social scientists, fisheries, and fish.

THE SOCIAL SCIENCE OF FISHING AND FISHERIES

When people in The Bahamas hear that I, an American academic, have spent time talking to fishers, their first response is to commend my work; they assume I am working to conserve marine resources by teaching fishers to recognize the threat they pose to the marine environment. They assume that people who make a living extracting natural resources will take anything and everything they can to make money, unless they possess crucial knowledge about scarcity. Another common assumption is that the marine management community needs to learn more about fishing activities to understand the extent of fishing impacts on the marine environment to make appropriate interventionist policy. My work is appreciated only if it can educate both managers and fishers about anthropogenic change. I have become an intermediary.

This middle-ground positioning is common in the fields of maritime and fisheries anthropology. Maritime anthropology has historically been associated with investigations of the lives of people who live by and off of the sea, and I discuss it here to show how the field has shaped understandings of fisheries in general so that we might follow those understandings as they evolve in the Anthropocene. Much of the work done in this vein has been concerned with collecting fisher narratives and classifying fishing economies and forms of social adaptation and organization around fishing practices.[8] This field exemplifies the way that the sciences of conservation and management and the social sciences are becoming increasingly interconnected within Anthropocene science.[9]

Fishing is of course a central concept for most maritime anthropologists, described as a productive activity that depends more on natural or wild processes than manufactured processes.[10] Fishing is therefore sometimes construed as closer to nature in this analytic tradition. Maritime anthropologists are "explicitly or implicitly concerned with what there is about a wet and fishy productive regime that defines the social, cultural, and economic life of fishing communities."[11]

As is evidenced by the DMR's management strategy to overfish the invasive, the lionfish and Bahamian fishers exist in an institutional world wherein nature is in crisis and their existence as appropriate parts of that nature is now in question. This framing legitimates efforts to manage people who fish, reflecting prevailing trends in marine social science. Maritime anthropology has therefore become most widely relevant as a tool for social impact analysis. As a form of applied anthropology, social impact analysis assesses the impacts of policy changes and major events on communities, groups, families, and individuals and is often mandated by law as an aspect of management projects involving drastic social change and intervention.[12] Anthropologists and other social scientists are increasingly called to complete these assessments, with this sort of collaboration becoming a matter of course. If "anthropologising on fishermen has become quite an industry," then maritime anthropology has increasingly become a form of policy-oriented anthropology.[13] It has become an example of a "support science": a social science that bolsters the designs of the natural sciences and helps to produce the socioecologics of the Anthropocene in ways that contribute to a certain style of management.[14]

In The Bahamas this bolstering use of social science in marine management is evidenced most clearly in studies of the impacts of marine protected areas (MPAs), especially since the country is home to the oldest MPA in the world, the Exuma Cays Land and Sea Park, created in the 1950s.[15] Still in use today, this park model separates fishers from sites of fish to protect both the fish and marine-based livelihoods from regional overfishing, and it has been rearticulated in the country's Marine Reserve Network, a chain of MPAs legislated throughout the archipelago.[16] It is generally agreed in the management community that commercial and artisanal fishing negatively affects marine reef biodiversity in the Caribbean and that protected area management must target fishing communities adjacent to planned MPAs for regulation.[17] But social scientists are rarely, if ever, called on to determine whether fishing communities are the appropriate sites of such scrutiny prior to their deployment in those communities.

The story of the lionfish both deviates from and complements the conceptual foundations of maritime anthropology.[18] I focus here on the concepts engendered by GCS-based notions of fisheries management that affect the way life in the sea is characterized in The Bahamas and in general.[19] Fisheries are an important entry into these changing relationships, and yet, despite the fact that fisheries are concrete objects of management, a fishery is still difficult to define. The word *fishery* encompasses the occupation of catching fish, the organized industry of catching fish, the season for catching certain fish, a place for catching fish, a fishing establishment, the legal right to take fish in a certain place, and the technology used to catch certain fish. All these meanings are situational, but what is striking in these definitions, and what is often overlooked in the policy-making process, is that a fishery is decidedly *not* a "natural" object, no matter how often it is construed as simply consisting of fish species. A fishery cannot be defined solely by what is said to be nonhuman or natural; rather, it is defined socially by what certain people can or might do to remove fish from the water, by what people consider to be "fish" in the first place, and by locations where it is or is not appropriate to fish. In other words, socially defining who, what, where, and when makes a fishery. Fishing, then, becomes an equally nebulous activity, and fishers become a part of an entire creative and political process involving the making of anthropogenic fisheries, a historical and social process of calling them into being and of separating them from what is said to be cultural. Therefore, fisheries are not merely dependent on ecology—they are dependent on design.

Fisheries are said to be about many things in the Anthropocene Islands: from the Sea Gardens of the late nineteenth century to a manageable natural entity, a vulnerable ecosystem, a regional database, a site of commercial or subsistence production, a frame for conceiving of socioecological systems, a global model, or the locus for the planetary collapse of marine biota. It is crucial to think about the *work* these various notions perform. Thinking critically about the form fisheries concepts take and the ways in which something like "the Bahamian fishery" has been designed or redesigned as a site of contestation in the Anthropocene should be an essential aspect of the social science of fisheries.

I suspect that, prior to its focus on environmental management, conservation, and fishing communities, marine anthropology was concerned with economic development and fishing communities. Along with small islands, current characterizations of fisheries, fish, and fishers that animate the lionfish story in The Bahamas likely find aspects of their contemporary form in international development, and for decades anthropologists have been involved in defining and designing these figures along with scientists, economists, and development practitioners.[20] Fishing is a development enterprise imbued with certain notions of fishers as possessing cultural values and recognizing economic value, with notions of fishable species having value as potential products in an international market, and notions of fisheries being the site of social values and market value amenable to capitalism or commercialism. Applied social studies of the social impacts of development planning have been historically used to produce figures that are integratable into development plans, upholding particular visions of value(s), fisheries, and fishers that uphold predetermined figures enrolled in studies and fisheries markets.

Moving from economic development to sustainable development in the Anthropocene, another example of fisheries design comes from the history of federal marine management in the United States. The National Marine Fisheries Service hired its first anthropologist in 1974, following the development of the notion of maximum sustainable yield and the idea that this notion required that "we would manage people—fish don't listen to you."[21] Sociocultural analysis of the community impacts of fisheries policy was partnered with economic analysis of natural science–based fisheries conservation projects, and social science was purposefully *integrated* into natural science. Fisheries management is now faced with challenges construed as the decline in marine resources and changes in marine environments owing to Anthropocene factors such as climate change, and these challenges are construed as affecting fishing communities and the relationship between community identity and the environment.[22] The feeling is that social science will become ever more necessary as the complex problems of the Anthropocene develop, involving more management pressures and fisheries in crisis.

Applied anthropologists have been instrumental, at least in the United States, in influencing the figures of institutionalized fishing and making what was understood as natural more cultural and socioecological and more approachable through social science. These same strategies are influential in The Bahamas, though in my experience social scientists involved in marine management are almost always foreign researchers, students, or consultants. The universal unit of anthropological analysis in this arena has become the community, organized around a similar sort of wet and fishy cultural ecology and susceptible to disruption by environmental and regulatory change. The notion of a sustainably managed and socially responsible fishery necessitates social impact analysis, requiring expertise that can demonstrate a specific kind of sociality for these natural systems in crisis and legitimate fishing communities as the site of targeted management.[23]

CONVERSATIONS ABOUT THREATENING MARINE SPECIES

Part of what now anchors understandings of both these species, fishers and lionfish, is the play of nature and culture in conversations about anthropogenic impacts. Fishers are said to negatively impact the marine environment by taking too much, while lionfish compound this problem as a consequence of unintended anthropogenic species introduction. These figures of dangerous, human-made change are thought to disrupt the natural fishy order. In the Anthropocene Islands of The Bahamas, the growth of social scientific categories that are amenable to policy interventions is amplified by growing international concern with the impacts of overfishing.

Overfishing is described as a global threat to most marine systems, especially fragile reef ecosystems like those in The Bahamas. There have been many instances of what managers recognize as overfishing that I witnessed during my time in The Bahamas. When our student research group was served fresh spiny-lobster tails out of season in the home of an Abaco Island resident as a welcoming gesture, we knew that they represented overfishing. When I watched one man pull into Nassau Harbor in a small Boston whaler weighted down under a massive pile of live queen conch, the algae-covered, undersized shells drying out rapidly in the hot sun, I knew that was overfishing. When an Eleutheran fishing family netted what seemed like an entire spawning aggregation of yellowtail snapper that died and rotted in the net due to lack of freezer space, the whole settlement knew that was overfishing. In each of these instances what was observed was illegal, a violation of Bahamian fisheries regulations. Yet viewing these examples with a unitary concept of overfishing or as merely evidence of economically pressured fishers driven to overfish misses something significant: the way in which these fished species and their fishers have been articulated as antagonistic opposites within contemporary fisheries management designs.

The conceptual package of the Bahamian fishery comes loaded with cultural notions about fishers, characterized most often in The Bahamas as "fishermen"; their communities; and their object of desire, fish. *Fish* denotes a large range of species including lobster (known as crawfish), conch, grouper, turtle, ray, sponge, snapper, dolphin fish, marlin, wahoo, crab, reef fish, and deep-sea fish. In the Anthropocene Islands both overfishing and invasive species are said to drive detrimental anthropogenic change leading to potential fisheries collapse. Therefore, the ways in which fishers and certain fish species are currently characterized as enemies within scientific fisheries studies and management designs directly determines policy recommendations and action plans in The Bahamas, the orchestration of nature and culture.

The marine research and management circles in The Bahamas are quite varied. The Ministry of Environment, the DMR, and the Bahamas Environment Science and Technology Commission are government agencies. The Bahamas National Trust is a quasi-governmental organization, funded by government and private interests. The Nature Conservancy of The Bahamas and the Bahamas Reef Environment Educational Foundation are both nongovernmental organizations with strong ties to international conservation organizations. Yet all these marine management organizations are primarily run by Bahamian staff, many of whom have higher degrees from the United States, Canada, or Europe, and the majority of the marine management guidelines come directly from international science and management arenas, especially when it comes to the implementation and design of MPAs.

Within these circles "Bahamian fishermen" come to wear many conceptual hats. Like the lionfish, they are compound creatures. They are considered to be a ubiquitous national group: "fishermen." Or they are described as discreet community units located in settlements throughout the country, rendered separate from the "Bahamian public" or "Bahamian society." For example, Spanish Wells, Cherokee Sound, and Tarpum Bay are all known as present-day fishing settlements. This diverse group is too often treated as a timeless social category: they are "fishermen," signifying a somewhat vague universal occupation.[24] At other times they are configured as opportunists whose exploitative activities increase with their own need. They are sometimes described as a specific "stakeholder" in the game of constantly negotiated fisheries management politics, a stakeholder that performs at certain times and is absent in others. At other times they are rendered knowable through the figure of the self-interested individual as explanation for human behavior. Maintaining the "livelihood" fishers are said to require has become a prime concern, though the appropriate form of this livelihood is in question.

In The Bahamas fishing was once considered a noble activity. It allowed people to make a living outside of the service economy and maintain a connection to the sea.[25] Just as I have witnessed instances of overfishing over the years, I have also experienced flourishing Bahamian lives with fish. I have watched many islanders on town docks and public waterfronts spending the early evening hours handlining for fish in the shallow waters. A hook and line without the rod is all that is needed to feel the vibration of a fish through your fingers and pull in a glimmering creature that will make a tasty meal in the frying pan. Bahamians of another socioeconomic status spend whole weekends in tournaments for large pelagic fish like dolphin, wahoo, and marlin or deep dropping for red snapper with golden eyes that bulge as they are brought to the surface from the depths.[26] I have purchased fish directly from commercial fishers at Potter's Cay Dock and the Montague Boat Ramp in Nassau, haggling over the price and quality of hogfish or grouper. These fishers will tell you with a sharp sense of humor tinged with earnest assurance that they have to charge the prices they do because the cost of gas is a burden, the work is laborious, and the hours are long. And many of them will also tell you that they are descended from families who have been fishing these waters for generations or that they apprenticed with fishers who had forty-year careers making a living from the sea.

But this reverent attitude is rare in management circles. Today fears of global overfishing are increasingly cited as the justification for fisheries studies, and fishers are more often construed by management as people who lack information about overfishing, information that experts possess and that could be passed to the fishers if the right channels could be found. They are not seen as possessors of specific and specialized local knowledge.[27] Overfishing is cited as something that has always happened in The Bahamas, and the general GSC attitude is that "humans have been disturbing marine ecosystems since they first learned how to fish."[28] This practice of "fishing down the food web" is attributed to most fishers, no matter what traits distinguish them, with little means for managers to determine actual responsibility or causal factors in specific situations.[29]

There tends to be little public discussion of the reasons behind the high demand for fisheries products or the fact that fishers seem to bear the brunt of fisheries conservation policies in the country. Instead, fishers have become major characters in fisheries dramas. I am interested in thinking about the ways in which varied characterizations of fishers allow for their portrayal as malleable entities, simultaneously transgressive agents of danger and strange sources for hope in a sustainable future designed around community—if only they could be informed.

Much like the Bahamian fishers, the lionfish has been perceived as a threatening figure in an ecosystem in possible crisis.[30] A 2009 invasive species poster created by the Bahamas National Trust has a large, dramatic photograph of a mature lionfish on a reef, the fish and its dramatic spines and fins taking up most of the frame. Another flyer made by the DMR has a photo of the lionfish, with the caption "Do not touch!!!"[31] The Nature Conservancy of The Bahamas website has a page devoted to the lionfish, stating, "Colorful tropical fish are fun to watch in an aquarium or home fish tank. But what happens when exotic fish are released into the wild—and start taking over the seas? . . . A popular aquarium species called the lionfish has been found in increasing numbers in The Bahamas, threatening to displace native fish and disrupt local fisheries."[32]

This notion of people disrupting a balanced ecosystem or natural environment has long been under attack, and the issue about what to do with ideas of the human and the natural are undergoing transformations in GCS circles. Ecologists and other scientists are attempting to reframe their language and their focus; new ecology, urban ecology, and restoration ecology are some disciplinary examples, as is the now familiar term *socioecology*, but these efforts in holistic thinking have been relatively unsuccessful in fisheries, as the human and the cultural still hold distinct conceptual positions from the ecological. And thus I consider the language of "human impacts" to be an impoverished way to think about ecological connection between human and nonhuman natures in the Anthropocene, with generic "humans" and ideas about "human behavior" consistently taking too much of the analytic focus.[33]

FIGURES IN EXCLUSIONARY ECOLOGIES

The lionfish captures headlines and redefines categories. Many scientists and managers fear that it has become "established," a dangerous condition in the language of anthropogenically induced invasion because this means it will be nearly impossible to eradicate. Anthropologists note that nature and culture have been "put into flux by the very idea of alien species" with the designation of invasive as the social judgment of harm.[34] Anthropologists also examine how ideas about what is native change over time and how biology has become an idiom for imagining cultural systems and the grounds for a notion of socially strategic nativeness.[35] In a sense, science produces ideas about nature and culture along with the reimagination of space and the strategic prescription of what should and should not be in it.

We can see this in the politics of invasion in the Anthropocene. In The Bahamas this politics hinges on the language of invaders and visitors, shifting distinctions of arrival and belonging. The very visible and dramatically venomous lionfish is figured as a transgressive anthropogenic enemy, highlighting conditions of existence in a Caribbean postcolonial nation-state.[36] Invasion is socially construed with the figure of the Haitian illegal migrant, often described on Bahamian national TV and public media as coming by water to suck up limited state welfare resources and weaken the capacity of the nation.[37] This figure of the invasive arrival, within which the lionfish fits all too nicely, is contrasted with that of the overly welcome visitor, the tourist and offshore finance investor who bring resources with them to be captured.[38]

Understanding lionfish beyond the reigning rhetoric of human impacts and the narrow politics of invasion requires an exploration of the conceptualization of animals that can shed new light on fish, fishers, and the ways they are implicated in contemporary fisheries management. The scientific logic that shapes nonhuman nature should become a focal point, and I would like to use three examples here to help rethink socioecological connections.[39] To start I consider the lionfish as an emergent "keystone species" for fisheries GCS research in The Bahamas.[40] Just like a biological keystone species are said to anchor an ecosystem, the presence of lionfish and the scientific fears about the animal's potential economic and ecological harm to the fishery anchors more scientific research in a postcolonial nation where scientists, foreign and domestic, often feel underappreciated and ignored. Even more than the fish, however, certain humans have been construed as key creatures in this fishery, and their stakeholder designation as fishers anchors and legitimates an interest in the social science of ecological systems for conservation practitioners and biologists.

Second, just as they are both keystone species, fishers and lionfish are also "gatekeepers."[41] In the socioecologics of Bahamian fisheries, they have been positioned as threats to the development of fisheries' value in reef biodiversity in that they are both construed as potentially overexploiting this resource. The proposed resolution for this problem, through the control of the lionfish population or the reeducation of fishers through outreach programs, forces a search for some indication from these creatures that these solutions are effective. The *presence* of lionfish and the *behavior* of fishers are recognized as the necessary indicators that hold these connections together and enable GCS research and management plans.

Finally, beyond mere representation, lionfish and fishers are produced in a concrete and corporeal sense through management practices. Lionfish and fishers, like other organisms, are actively "figured" through their interactions—they are formed into particular figures in the act of meeting. Human-animal collaborations, like fisheries, simultaneously create "creatures of imagined possibility and creatures of fierce ordinary reality."[42]

These three concepts, the keystone, the gatekeeper, and the process of figuration, help us understand that, like the celebrated sea turtle, a popular and "friendly" conservation icon and "ecofetish" in the Caribbean, both lionfish and fishers are figured as key species and gatekeepers that uphold the legitimacy of evolving fisheries management practices.[43] But, unlike the sea turtle, fishers and lionfish are managed as though they should be separated from healthily functioning Caribbean ecosystems. Rather than focusing on "mutual ecologies" of humans and animals creating new marine systems, these management practices have become more exclusionary when it comes to particular human and animal figures.[44] Acts of figuration around the lionfish involve the enumeration of fish and fisher characteristics and the investigation of their practices, shaping the way both fish and fishers are understood in The Bahamas while entrenching the authority of GCS management at the same time.

BOTH DANGEROUS AND DELICIOUS

Studying invasibility in The Bahamas—the problem that occupied the field experimentation of the graduate student I assisted—is not a new activity. In fact, the islands' natural history has always involved discussions of anthropogenic plant, mammal, bird, and amphibian invasion. A zoologist of the region has stated that "particularly since the arrival of Columbus in The Bahamas a half millennium ago, the Islands have been invaded by a host of animals and plants, some stowaways, others intentionally introduced."[45] Lionfish, as aquatic invaders, are only the most recent instantiation of this ongoing discussion of island vulnerability, invasion, and transgressive danger in what GCS researchers focused on invasive species construe as an era of global change.[46] The current scientific response to the fish can be summarized this way: "It would be prudent for affected nations to initiate targeted lionfish control efforts as soon as possible. Concerted and sustained efforts to reduce densities of lionfish at key locations, including potential 'choke' or dispersal points, as well as particularly vulnerable or valuable reef areas, may help to mitigate their ecological impacts."[47]

To tackle this synergistic problem of overfishing, the potential collapse of fisheries, and enemy species, fisheries management organizations in The Bahamas have moved to make the lionfish a legitimate part of the Bahamian fishery. But those in charge of managing the Bahamian fishery are not advocating the complete destruction of the species, which may prove impossible. Instead, they are designing means to internalize the fish into the realm of the even more familiar. As one marine biologist says, "Bahamian officials can institute a bounty and convince locals that lionfish taste like chicken and are easy to eat once you know how to avoid the spines and cook the fish sufficiently to denature the venom."[48] Along with the creation of a national lionfish database as part of the Bahamas National Lionfish Response Project, officials now advocate for the lionfish to become a commercial fish species as part of the viable Bahamian fishery, to be eaten at home and in tourist restaurants.[49]

In 2009 the Bahamas National Trust instituted the first Lionfish Control Project, wherein fishers were invited to enter in a two-day lionfish-catching tournament, with cash prizes given to the boat to catch the most lionfish, the largest lionfish, and the smallest lionfish. The tally, weigh-in, and prize distribution were held at a restaurant in Nassau, where lionfish were cleaned, cooked, and served as part of the event to demonstrate that they are edible and enjoyable. A Nassau daily paper covered the event, stating that "enthusiasm is growing for the fish as a source of food," and Michael Braynen, director of the DMR, said, "We think this is perhaps the best avenue that we will be able to pursue to reduce the numbers of lionfish in the environment by turning them into a fish that people see as a food source, that commercial fishermen will want to take and that people will want to buy."[50] It is remarkable that GCS management organizations explicitly designed this fish, allowing the species to become part of the national fishery, making it a model species that signifies the vulnerability and value of the Bahamian marine environment. The fish has become a "totem animal" for the Anthropocene.[51]

The materiality of this totem fish became very real for me in 2012. I heard through fishing networks that a high-end tourist restaurant at a Nassau megaresort had put lionfish on the menu, so I saved my money and went on a busy night during the tourist season. I discovered that locally caught lionfish had become a flashy and expensive sashimi dish. Sold then for forty-five dollars a plate, it was the only thing I could order. Instead of the endangered and guilt-riddled bluefin tuna, flown in

FIGURE 11. Lionfish tournament poster, Bimini. Woody Foundation (2014).

half-dead from the other side of the world, patrons could choose lionfish as a special local species, easy on both carbon footprint and morals. The fish arrived in many thin pieces on a platter, resplendent in its stripes even after death, accompanied by impressed comments from my companions and nearby diners. I took the first bite, and I had to admit that the lionfish was perfect. I also realized that in this luxury Japanese restaurant chain with branches around the world, only diners at the Bahama branch had the option to order a taste of the Anthropocene Islands.

I had another run-in with the lionfish at the fourth annual Lionfish Derby in Nassau in 2014. Sponsored by the Bahamas National Trust and the DMR, the derby took place one afternoon on another dock next to a casual, harbor-side bar and grill patronized by locals and boating tour-

FIGURE 12. Sashimi lionfish at a high-end Japanese restaurant, Nassau. Photo by author (2012).

ists. Participants registered for the derby by boat, leaving early in the morning to "murder fish," as derby fishers commonly say. By the midafternoon these boats pulled up to the derby dock in a steady stream, and tallying went on throughout the day. The lionfish were typically speared out on the reefs surrounding New Providence, then kept in coolers on board until an official counted, weighed, and measured them. Each boat's results were posted on a white board in the restaurant for all to see.

The Bahamas National Trust had a booth with T-shirts, a minibar, and brochures about the invasive lionfish threat. There was also a cleaning station where two volunteers—a recreational fisher and a DMR officer—demonstrated safe lionfish-cleaning techniques and answered questions from curious onlookers, primarily about the neurotoxin in lionfish spines. They mentioned often that the venom has never caused any reported deaths, but that it could severely hurt someone with heart defects. In any case, they explained, a sting requires medical attention because the pain is intense and steadily increases in the first two hours. In between serious demonstrations they joked about the wholesale slaughter of the species, saying, "dead fish don't breed."

FIGURE 13. Lionfish cleaning demonstration, Nassau. Photo by author (2014).

The fish heads and guts were thrown into the water, where they would eventually become food for sharks and stingrays, and some of the fillets were given to the host restaurant to serve to promote the event. On that day the bar and grill was serving "lionfish popcorn"—spicy, bite-sized balls of deep-fried lionfish served with a wedge of lime and island tartar sauce—and lionfish sliders. While this place was not nearly as fancy or expensive as the high-end Japanese restaurant, tourists and Bahamians were again encouraged to try the fish as a local "guilt-free" snack.

Control through consumption is a solution often posed to combat perceived Anthropocene problems, and just as with other Bahamian sustainable development concerns described in this book, tourism is a large part of the recommended form of consumption for lionfish management. But not all Bahamians approve of this institutionalized linkage. Tamika Galanis, a Bahamian artist and intellectual, has called the spread of lionfish in the Americas, "the largest marine disaster in history." Her short film, *When the Lionfish Came,* depicts lionfish as metaphors for the predatory and invasive spread of neoliberal capitalism in the Caribbean region that leads to dependency, debt, and the pervasive consumption of regional culture and identity through foreign-owned industries like tourism.[52] The message promoted to visitors through the national lionfish management plan ("While you are here, order lionfish") is a heavy irony for Galanis, in that the fish invasion itself was likely sparked by the release of lionfish from resort aquariums, requiring more tourism to clean up the mess made by tourism in the first place. The lionfish, in this final figuration, becomes yet another representative of the plantation economy that repeatedly re-creates and reconsumes Bahamian lives.

Examining the construction of seemingly matter-of-fact categories is a step toward rethinking them. Social science, attending to the strategic positioning of fish and fishers by the natural sciences in the Anthropocene, can show how the lionfish is implicated in more than an overfishing problem; it is also a problem of homogeneous conceptualization, even in supposedly "complex systems." While the lionfish has become an Anthropocene totem animal for GCS, the "threatening" status of Bahamian fishers has become even more entrenched in some ways as a result. Despite evidence that not all fishing is necessarily a direct cause of the declining health of the marine environment, evidence that no-take protected areas are not necessarily effective for conserving dynamic ecosystems, and evidence that targeting fishers for regulation causes them to suffer the majority of the penalties while shouldering the greatest risks, especially when compared to the consumptive tourism industry, many still perceive fishers as overzealous consumers ("fish murderers") in fragile marine environments.[53]

The thinking that pits fishers against fish to save the fishery—the story of the lionfish—internalizes the malleable figures of fishers and lionfish into the Anthropocene Islands. The lionfish is no longer perceived as only the enemy invader; it is becoming internalized into the fishery as a commodity species. The strategy to have fishers fish for this species and serve it to islanders and their paying guests will supposedly succeed precisely because Bahamian fishers are imagined to be transgressive overfishers. Thus, by "doing what they do best," fishers will also supposedly sustain both the fishery and their own livelihoods as they are internalized into lionfish management plans. The Anthropocenic assumptions about impact—of the detrimental human impacts on the environment and the detrimental social impacts of management planning—become rearticulated in this scenario.

Yet the stakes are higher for some than others. Despite the development of cutting-edge GCS notions of complexity, such as networked socioecological systems, small-scale fishers in The Bahamas are entangled in the simplifying figurations of international marine conservation discourse and policy that puts too little emphasis on wider consumption trends in favor of targeted local interventions.[54] And blame, responsibility, and behavioral research vectors continue to be aimed at those who are less likely to be involved at higher levels of policy making for Anthropocene issues. These vectors complicate the evolving role of social science in GCS framed, interdisciplinary fisheries studies.

Without these notions of human impacts and these particular framings of nature and culture, fishers and lionfish would not be such significant anthropogenic figures used to maintain GCS management as a strategic arena of design and action. Further, the way in which the nonhuman fish is figured in this domain also implicates human fishers, because figuration is a relational process. My hope is that it will become increasingly obvious that the creation of a fishery is first and foremost a world-making practice for the Anthropocene, not just the discovery of a vulnerable entity in automatic need of exclusionary protection or sustainable consumption.

I have presented a case for recognizing the lionfish as an iconic species for the Anthropocene. But this totem is significant not only because of its anthropogenic origins in the West and its potential for spreading anthropogenic doom in sensitive marine ecosystems. The lionfish is also a significant animal in that it now helps to link fisheries and tourism throughout its invasive range, sparking the Anthropocene irony accompanying the claim that this era of global change is also an opportunity for those in the best position to capitalize on such change.

The malleable figures of fishers and lionfish are designed together by GCS-based fisheries science and policy. This technocratic realm increasingly relies on the support of applied anthropology, on notions of outreach and ideas of sustainability, on the language of livelihood, and on community-based socioecology and the shifting relations of nature and culture construed therein. If "every fish that gets caught is partly that of others" owing to the social relations of knowledge and practice that create and are created through fishing, then the lionfish is in the process of socially becoming a fishery, part of a fishery, and a fish, in the sense that it can be fished.[55] This may represent the exploitation of visitors, a theme from Caribbean tourism, but the lionfish has also become a representative figure for the notion that fisheries are designed through the creativity of fisheries management in the Anthropocene, something that anthropologists can track from their position within the sciences.[56] These fish have been designed to be both threatening and redemptive, both dangerous and delicious, for Bahamian communities, ecologies, tourist economies, and contemporary management experts.

5

Down the Blue Hole

GOING TO EXTREMES AND BACK AGAIN

Sometimes you go and it's black, black, black. Now, why is that? You all should know, you're scientists. Then you go and the water is clear. Beautiful. You can see the fish swimming. You might even chance a swim [laughs]. I don't know why it changes like that, just depends on the day, I guess. Maybe the rain.

—Androsian woman, discussing Gobbler Hole

We need people to understand the reason behind blue hole protection so they can take ownership of their history.

—AMMC representative

I have known of the existence of blue holes since my very first visit to The Bahamas, well over a decade ago.[1] One slow summer day a Bahamian fisher took me by boat out into the wetlands surrounding Cherokee Sound. He took care to point out large dark holes in the sea floor as we cruised slowly down the mangrove channels. This fisher friend frightened me when he told me that the holes boil and breathe, connected to the tidal action of the surrounding sea; they can suck swimmers in and drown them. A week or so later on that same visit, my research supervisor took me out to act as lookout while he dove with scuba gear into a lonely blue hole in the middle of the mangroves. "If I don't come back in forty-five minutes," he said, "drive this boat back to town for help." He seemed unconcerned that I did not know how to operate the outboard motor on our small dingy. And again, a few years

FIGURE 14. Tanks and transmitters at the edge of a blue hole, Abaco. Photo by author (2016).

later in Rock Sound, a settlement in southern Eleuthera, I was shown another blue hole. This one was in the middle of town, resembling a small lake with vertical rocky banks. Residents of the settlement had carved stone steps down one side to allow for safe swimming. This hole was said to be bottomless, but it was full of ocean fish—even grouper—so townspeople knew it was somehow connected to the sea. Later I was told that this blue hole was not bottomless, and that old cars and refrigerators had been salvaged from its depths in periodic cleanup efforts.

Blue holes transform when they travel outside of The Bahamas. In February 2010 the U.S. Public Broadcasting Station aired a prime-time NOVA program, titled *Extreme Cave Diving*, introducing U.S. viewers to Bahamian blue holes and the scientific exploration done within them. The show defined blue holes as "underwater caves that formed during the last ice age, when sea level was nearly 400 feet below what it is today," and they were described as "little-known treasures of the Bahamas" and as "one of Earth's least explored and perhaps most dangerous frontiers."[2] An interdisciplinary team of researchers, including an anthropologist, photographer, microbiologist, paleoecologist, underwater cave explorer, and a Bahamian archaeologist were described as "astronauts of inner-space," delving into a Bahamian attraction well outside the familiar realm of sun, sand, and sea.[3] Blue holes were figured as vitally important because they acted as time capsules: the deepest layers of the caves are anoxic, preserving the remains of whatever creatures, human and nonhuman, might have fallen into these aquatic tar pits over the centuries.

Blue holes have particular speleological characteristics, and they are of particular interest to those who see caves as "natural laboratories."[4] The TV show is an example of this kind of global change science (GCS) blue hole narrative about The Bahamas: that The Bahamas was once a "thriving Eden" of diverse species and verdant plant growth, though today the islands are arid and sparsely inhabited by terrestrial creatures. One aspect of this cave exploration is to bring up evidence that can explain the shift in island life from the "lost world" of the past to the present day. Another aspect is the search for fossil remains of new species that have not yet been cataloged by science.

To date, the remains of an ancient tortoise, crocodile, lizard, owl, boa, and flightless bird species have already been discovered. Thirty-three are said to be species that were not known to have inhabited these islands, and three are said to be new to science. The dive team has also discovered human remains in the holes, possibly of Lucayan origin, dated to be eight hundred years old, and this has sparked speculation that the die-off of species was related to the presence of humans in the islands.[5]

The Bahama Island chain lies near the Gulf Stream in the Atlantic Ocean, a movement of oceanic water that has important implications for the planet's climate cycling. Stalagmites in the blue holes—long spears of limestone that hang down into these underwater cavities—hold a microscopic record of the ancient climate and the activities of the Atlantic Gyre. These spears can be read much like glacial ice cores in other parts of the world to mine for climate history and determine how fast the ancient climate changed. These aspects of the holes were framed in the show as "the secrets of the planet's history."

Blue hole exploration is more dangerous than any other kind of marine archaeology because of the lack of vision due to darkness and silt; the depths, which can cause nitrogen buildup in the body; the narrowness of the cave passages; the unknown length and end point of many of the caves; and the prevalence of cave-ins. In the NOVA special the underwater photographer even found the remains of an unnamed and long-dead cave diver in one of the holes they filmed. In the face of this extreme milieu, cave divers are characterized as intrepid and brave explorer experts, risking their lives in the name of science. In this example they were filmed in their dive gear, covered head to toe in wet suits, tubes, rebreathers, weight belts, fins, and masks.

The previous description is one that suits the classic exploration and discovery theme of many scientific narratives that ground the well-known colonial figure of the adventurer scientist.[6] However, my experience with blue hole research in The Bahamas complicates this narrative. Blue holes can reveal the ways in which the GCS field sciences contribute to the Caribbean project of national identity formation in the Anthropocene through the development of science-based narratives that make new connections between pasts, presents, and futures. Blue holes are also another example of the process of developing alternative sites for the expansion of tourism—literally designing new places and practices of visitation that focus the industry's attention on socioecology as a new arena to explore.

The examples presented here come from my experience as a visiting researcher in Abaco and Andros, two large islands in the northern Bahamas. In Abaco I became a participant-observer in a performance of "science history" involving blue holes, and my labor was a necessary part of the show. In Andros, several months later, I was a social scientist responsible for collecting stories and memories from islanders. Through my engagement with these projects, I learned how the mysterious—the deep, the dark, the forgotten, the unknown—can be brought into the harsh light of the sun and the field lab and eventually onto the itinerary of the island tour, should events occur as some participants hope they might.

Once interesting because some possess a "controlled" physical environment like that of a laboratory, blue holes are now also interesting because of their unique social as well as physical "placeness." They are another example of the field as an important site of GCS knowledge production. Away from the TV screen, in The Bahamas itself, these mysterious blue holes are in the process of being made familiar, preservable, and ripe for visitation, to stand as a symbol of vital planetary knowledge about anthropogenic change but also as sites of national heritage with possible applications for "heritage tourism."

Heritage is a recent industry buzzword marking the diversification of tourism into the market for regional and global history and the desire for destinations to stand out in a sea of beach offerings. In fact, The Bahamas ratified the UNESCO World Heritage Convention in 2014 and has already proposed sites for the World Heritage List, including the lighthouse system of the archipelago and Inagua National Park.[7] The blue holes of The Bahamas are now also under consideration as possible World Heritage sites.

Heritage is considered a new arena in which to benefit the "tourism product" of The Bahamas because of the potential to attract a whole new range of educated and adventurous visitors. But this product must first be developed in an accessible and understandable way, and the knowledge produced by transnational GCS research about features like blue holes will play an integral role in this development. Marketing Bahamian geology through the creation of GCS-informed features falls within the designation of heritage tourism and within the purview of Bahamian institutions such as the Antiquities Monuments and Museums Corporation (AMMC). The examples here reveal the latest plans to remake Bahamian geologic formations into features for Anthropocene travel markets.

ABACO

The northern Bahama Islands are sticky and warm in December. The afternoon sun feels intense, the winds have disappeared, and the long shadows cast by the spindly pines provide insufficient shade for sweaty, tired bodies. In Abaco the pines grow together in curious forests, sticking straight as a ship's mast out of the patches of dry underbrush, golden grasses, and rocky ground. In some places the trees seem evenly placed, as though planted so as not to touch or overshadow any of the others, and there is the illusion that one can see through the forest for miles. If there were only a breeze it would flow freely through the open air between the trunks.

In December 2008 I went to the pine forests of central Abaco to participate in the film project that would eventually become the NOVA special, and while I went in the capacity of an anthropologist investigating the production, I found myself quickly and summarily assigned as a sound technician, filling an empty role. At the time I knew only that the project was centered around recent discoveries made in several of the island's blue holes—what I understood to be water-filled sink holes whose depth, inaccessibility, and biological characteristics make them perfect repositories for all manner of interesting objects. I was just one of a number of participants in what was a consciously interdisciplinary endeavor, the kind of project I would later identify as exemplary of GCS. There were a number of U.S. scientists in attendance who did not make it into the television special, including a marine cave zoologist and a cave microbiologist; there was the U.S. film crew, which consisted of the director, a young camera operator, and suddenly myself; and there were the hosts and coordinators of the project: a Bahamian paleontologist from a local environmental nongovernmental organization based in Abaco and an American anthropologist who helped organize and mobilize the whole operation. Finally, there was a Bahamian marine biology student in attendance, back from her U.S. university for this project.

My time with the film production was exhausting, but not unpleasant. I list the players here and in the next section in an impersonal manner to protect their identity, but I don't wish to render them as automatons, identifiable only through their disciplines or through the loaded and obfuscating designation of "scientist." Everyone was involved because they had experience in the region or because they were interested in interdisciplinary work on caves. The shoot was permeated with a sense of adventure and exploration, a sense that had to do with the shared nature of the experience as much as with the nature being investigated. We were developing a more holistic design for producing knowledge about blue holes.

Blue holes are technically either submerged holes in the ocean, resembling deep blue orifices in the sea floor, or terrestrial holes filled with water, resembling a dark inland lake or pond.[8] The work I participated in involved mostly terrestrial blue holes, but blue holes in The Bahamas, terrestrial or marine, are generally formed by the same geologic "cave formation recipe."[9] Blue holes provide opportunities for speculation in the Anthropocene about the potential bioprospecting of natural marine products from cave-adapted animals, the secrets of climate change hidden in their preserved carbon, the life cycles of extinct plant and animal life, the chemical conditions of the early planet Earth, and the past migrations of people through the islands and their ability to change ecosystems. Standing on the edge of a deep and lonely terrestrial blue hole in the pine forests of Abaco, accompanied by a team of experts preparing to probe the depths with their bodies and an array of other instruments, I could not help but feel that there was something special at work.

Blue holes are now in vogue in The Bahamas and in Abaco among visiting scientists but also increasingly among regional and domestic scientists. This is in part because over a year prior to the convening of this film crew, another collaboration of sorts had taken place in Abaco at a blue hole called Sawmill Sink, an event that proved to be a quiet revelation to the scientific community of The Bahamas and to regional scientists in general. Exploratory cave dives in Sawmill Sink revealed a trove of archaeological and paleontological remains, including the fossilized bones of ancient birds, bats, turtles, crocodiles, rodents, and humans, as well as plant fossils, all remarkably preserved in the anoxic saltwater layer within the sink hole. The scientists involved in the exploration of this blue hole have begun to tell a particular story about the pre-Columbian Bahamas and the Caribbean, claiming that Sawmill Sink is the first area in the West Indies where such a collection of fossils has been found in such good condition.

From this fossil record scientists have been able to determine that a species of terrestrial crocodile and a giant tortoise once lived in The Bahama Islands, going extinct, they suppose, after the arrival of predatory human populations. They have also been able to determine that the climate and island habitat has changed drastically, once allowing for open grasslands where now there are the Bahamian pine forests. An American herpetologist involved in the Sawmill Sink excavations situated the relevance of the discoveries for science by saying, "The fossils from Sawmill Sink open up unparalleled opportunities for doing much more sophisticated work than ever before in reconstructing the ancient plant and animal communities of The Bahamas. It helps us to understand not only how individual species evolve on islands, but how these communities changed with the arrival of people because we know that changes in the ecosystem are much more dramatic on islands than they are on continents."[10] The film project was proposed after these discoveries were made in Sawmill Sink.

While struggling to operate the sound equipment, including the unwieldy boom microphone, I began to realize that what we were all participating in was a demonstration of Anthropocene socioecologics. Out in the pine forest, at the edges of a deep, water-filled hole patrolled by dragonflies and swallows, the divers and the scientists, and their copious equipment, the piles of wires and the plastic tubs of tools, were all performing for a film about the interdisciplinary science of blue holes and the excitement of making discoveries with global significance for the understanding of anthropogenesis.

This notion was further spelled out for me when the director of the Bahamian AMMC arrived in Abaco to visit our film site and make a statement on camera with a blue hole in the background. Established by government in 1998, the AMMC is a quasi-governmental organization whose mandate is to preserve the material and cultural heritage of The Bahamas to maintain and develop a sense of national history and to make this heritage "have a direct impact on the economy of The Bahamas" by opening it to tourism.[11] Some things construed as environmental and natural get caught up in this mandate in interesting ways because the AMMC is a body with cojurisdiction over certain protected areas that contain archaeological, cultural, paleontological, or biological artifacts.

On camera, in a brief and staged conversation with a member of the project team, the director stated that blue holes represent the "science history" of The Bahamas and the world and that blue hole scientific research develops this national and global history. He further noted that interdisciplinary international research teams can participate in capacity building, networking, and setting an example within the country and that work such as this can help show that it is "imperative to define government ownership of these important blue holes so as to avoid disputes and guarantee preservation." Then, to help cement this collaborative and productive relationship between the scientists and the AMMC, a new species of cave shrimp found in this particular hole by the team zoologist was presented to the director, to be named after him.

Straining to capture all that was said with the heavy microphone, I witnessed these unabashed political moves to make Bahamian blue holes simultaneously national and global, to tie the AMMC and the Bahamian government to future blue hole research, to inspire the formal state protection of Bahamian blue holes, to demonstrate collaboration with Bahamian authorities, and to thank the director for participating in the film project. In this performance both blue holes and cave shrimp were tied to the construction of a Bahamian socioecological history in which The Bahamas becomes suddenly relevant on the scale of evolutionary and geologic time. This history is mediated by GCS and the disciplines that design and define the Anthropocene. In other words, we were performing the production of a scientific, regional, and planetary history designed for use as particularly Bahamian national history. The significance of this history and its implications for heritage tourism were not clear to me until I did some further work with blue holes the following year. This work was deeply contrastive with my participation in Abaco, in that it required that I take on a major role as a social scientist and participate in the development of a kind of cultural history.

ANDROS

In June 2009 I returned to The Bahamas to help colead a group of Bahamian students on an ethnographic exploration of social life with blue holes in Andros, another island in the northern Bahamas. Andros, the largest island in the Bahamian archipelago and the island with the largest stores of freshwater, is blue hole country. With well over a hundred documented terrestrial blue holes, Andros may have more blue holes than any other location in the world.[12]

The same loose collaboration that gave rise to the NOVA blue hole film project in Abaco sponsored my participation in Andros. This additional work was intended to fill the gaps in the interdisciplinary data that had been collected in the holes thus far. There had been a marked absence of social research about blue holes to accompany the scientific investigations, and the project creators felt that the interdisciplinary rationale behind the project necessitated some sort of social inquiry. I therefore went to Andros, along with another American graduate student and a faculty member from the College of The Bahamas, to pilot ethnographic methods for the study of blue holes and to help make this project available as an educational opportunity for Bahamian college students.

Andros in June is decidedly not Abaco in December. June is the rainy season, and that June was especially rainy, characterized by thunderstorms and great gray masses of cloud. The pine forests of Andros were choked with damp underbrush, scuttling land crabs scavenged in dead pine needles and rocky holes, and buzzards circled continually or perched on the edge of vision, ruffling wet feathers.[13] There was no filming on this trip, no crew, no natural scientists, and our focus was resolutely on the four students, their development as researchers, and their ability to collect viable social data. We asked them to take an open-ended set of questions into interviews with residents of settlements in central and northern Andros. The interview questions asked about names, memories, and stories of blue holes; ideas about their creation and current relevance; and the past and present uses of blue holes for those people who had grown up in their vicinity and for those whose families had been there for generations. We asked that the students privilege elderly people or people who had lived on the island for most of their lives.

We heard many stories delineating the fear of blue holes for people who live around them. In a region where a number of people who reside by the sea do not know how to swim, many people in that part of Andros have lost relatives and friends to the holes, their bodies swallowed in pits that some describe as "bottomless." The holes are also filled with mermaids—harsh creatures who can drag a person down and hold them in the depths for years or drown them to exact revenge for the murder of fish.[14] But we also heard stories from people who used the holes as sites for swimming and diving, for washing clothes in their youth, and for catching land crabs, a regional delicacy and Androsian staple. We learned that taxi drivers already take visitors to the holes to swim and that they have been increasing in popularity for the past few years, as visitors become more adventurous and as dive operations make them more accessible. All these things and more we documented with audio recorders, interview spreadsheets, and student notes.

We were, in effect, creating the beginnings of a social history of a geographic feature for the Anthropocene Islands. We were demonstrating the social life of a "natural" form by documenting its interrelations with island people, looking for changes over time and between generations. As anthropologists and social scientists, my research partners and I were lending our particular form of expertise to this endeavor of producing a science-based understanding of the global and national importance of these mysterious holes. As part of the interdisciplinary project, we designed a template for assessing blue hole sociality in the rest of the country. Our questions are transferable, purposefully so, and if funding is found, as some hope it will be, more students will interview more people on more islands about the presence of blue holes in their lives, to create a national environmental history.

This social work in Andros aligns with the AMMC's mandate to make its designated resources and monuments "impact" the Bahamian economy in ways that the NOVA special cannot. By developing a cultural history of blue holes in Andros, we were creating a narrative that, when interwoven with the natural scientific research done by the team, would animate the "science history" of blue holes that could make them a specifically Bahamian geographic feature for the local "tourism product" in a travel market for adventure based on anthropogenesis. Contemporary blue hole productions about national history in The Bahamas led me to reexamine foundational Caribbean questions.

RE-CREATING CARIBBEAN IDENTITY AND TOURIST LANDSCAPES

GCS narratives about island evolution "discovered" inside blue holes—the Bahamian "science history" the AMMC director mentioned in Abaco—contribute to the process of nation building and identity formation in The Bahamas. This is also an ideological project that is of special concern for the tourism industry in the region. Not coincidentally, the question of national history and identity was a central concern in Caribbean studies in the late twentieth century, with much of the focus on the development of nationalist historical narratives.[15] Therefore, understanding the full significance of blue hole research for The Bahamas requires exploring a few examples from the Caribbean.

In Trinidad, another island nation formerly colonized by the British, engendering patriotism as a form of postcolonial self-consciousness required imagining the nation as having ancient indigenous roots. Individual claims to Carib heritage anchored Trinidad's history in deep time.[16] The discoveries in the blue holes of The Bahamas are similarly about imagining ancient roots. Blue hole GCS creates new national narratives based in ideas of biological nature that are imagined to be foundational and authentic. Thus, "science history" is rooted in current understandings of the relationship between ancient and contemporary humans, nonhuman organisms, and their environments. In Bahamian blue holes ancient Lucayan remains are not directly linked to any existing populations of people within the islands, because of the "postcolonial erasure of primordiality" for the people of the region.[17] Instead, these remains have been used to recount a story of destruction involving anthropogenically driven island extinction in which the presence of humans changed the island ecology forever.

Saint Vincent is known for having a homogenous national identity relative to some other Caribbean nations.[18] For some scholars this island exemplifies the way that creole models of a unified society have largely won out in some majority Afro-Caribbean middle-class nations with a shared history of slavery and black-white encounter.[19] Postcolonial national identity then stems from this history of shared creole culture and historical narrative. Some have argued that The Bahamas is also a nation with a similarly homogenized Afro-European popular historical narrative, but the GCS productions around blue holes may create new national, individual, and collective narratives based on revised knowledge about human life in these islands.[20] It remains to be seen how the natural history discovered down the holes will play out in terms of individual self-knowledge or a sense of collective nationalism. At the very least GCS socioecologics in The Bahamas—including blue hole research—provide alternative narratives for national identification with place in the Anthropocene.[21]

In contrast to these discussions of specific national historical narratives, Caribbean scholars also think about the conditions that shape Caribbean regional identity.[22] Caribbean identities result from the way in which people are positioned by historical events and the way in which they position themselves within historical narrative. In other words, they do not spring from any foundational cultural essence. Emergent technologies can facilitate such positioning.

Caribbeanist scholarship reveals the cultural implications of blue hole GCS and scientific narratives about the historical rootedness of geologic features. These are narratives about place, space, and nature that are crucial for an understanding of the contemporary region in an era of anthropogenic planetary consciousness. If we question the political desire for a Caribbean identity or even a specifically Bahamian national identity, then we can see that one aspect of the development of blue holes as a science-based feature might be a response to pressures from the tourist industry to redevelop the Bahamian "tourism product" and the national identity branded therein.[23] GCS narratives globalize the significance of blue hole nature while simultaneously creating a particular sense of valuable local heritage, and this is exactly how tourism branding functions.

I see the scientific exploration of blue holes as a potential tourist product in formation, a renarration of history, and a site of socioecological knowledge production—all based on the manipulation of space and place. Rather than focusing solely on the scientific "truths" found within blue holes or on their utility for re-creating a stable Bahamian identity, blue hole research should also point our attention to the role tourism plays in combining both of these processes to remake Bahamian landscapes. For many scholars tourism is a particular form of social relation that is deeply rooted in the commodification of place to match tourist expectations and desires.[24] But the blue hole example complicates the study of tourism. Instead, it is the *designers* of tourist products and the industry officials who envision tourist populations that challenge "the tourist" as the primary driver of change in the tourism destination of The Bahamas.[25] Tourism industry practitioners actively create new markets, collaborations, aesthetics, brands, and social practices, and the involvement of the AMMC in the blue hole film project is but one example of practitioners at work.

Bahamian blue holes reveal a proliferation of ways in which to think about the interrelation between scientific knowledge and island industry; international GCS and global, regional, or national narratives; and the nature of Caribbean tourist markets and markets for particular forms of nature in the Anthropocene. Blue holes are precisely *not* preexisting, primordial, or static landscapes. What is interesting about blue holes is that they are Anthropocene landscapes that are being actively *re-created in the present* as sites for ongoing scientific discovery and the perpetual revision of the social relations of tourism.

We can see the process of re-creation at work. In Abaco and Andros there was a continual interplay of national, regional, and global specificity and generality in blue hole research and exploration. The GCS projects practiced in and around these holes were largely conducted by foreign experts, with important exceptions, but this work was done under the auspices of several Bahamian institutions whose interest in the project and relationship to the holes is predicated on trying to develop a sense of globally significant locality and accessible "placeness." The development of landscapes of significance is a large part of the tourist industry's evolving relationship with the field sciences in the Anthropocene Islands.

The blue hole projects I describe here are material and symbolic processes of reimagination and re-creation. They represent a mode of apprehending these mysterious caverns wherein the microbiological processes of deep time, the ancient cycling of island ecosystems and changes in climate, the evolution and extinction of island organisms, and the human history of the region come together to legitimate the creation of blue hole features under the auspices of Bahamian heritage and global socioecological importance. Within such Anthropocene logics, moves to protect particular landscapes hinge on design, product development, perceptions of anthropogenic vulnerability, and the politics of identity and subjectivity.

MAKING THE UNDERGROUND MATTER IN THE ANTHROPOCENE

GCS field researchers re-create spaces through the experimental manipulation and development of scientific tools designed to "read how nature works."[26] These practices have a long history. Colonial cartographers, surveyors, botanists, natural historians, and the like dissected their surroundings, enumerating, charting, and depicting organisms, spaces, objects, and people, attempting to legitimate the reach of rule based on specific ways of knowing the world. Postcolonial nation building continues these practices, basing knowledge of the nation on what can be *made known* about the nation—on what can be scientifically demonstrated. The enlistment of GCS field science in the creation of Anthropocene space is, at its core, another means to legitimate claims to territory, resources, and ideas about life and collectivity.

One thing blue holes render tangible in this process is the fact that the life sciences and the social sciences are bedfellows when it comes to redesigning space for the Anthropocene. My experience with interdisciplinary blue hole investigations in The Bahamas makes this interrelation explicit. The combination of the expertise and authority of a number of field disciplines in one project—from marine zoology to anthropology—creates a holistic vision of blue holes as valuable and potentially visitable features of Bahamian socioecological history. It is hoped that they may one day materialize a Bahamianized global and regional identity as part of the AMMC's mandate to develop heritage destinations.

The Bahamas is of course not the only place in the world where archaeological, ecological, and geologic features are developed and exploited for the international tourist industry, but there are few parts of the world where tourism is so pervasive and so central to the economic viability, identity, and sovereignty of a nation.[27] This is in large part why my emphasis here has been on the "heritage" interests of the tourism industry, embodied by the AMMC, which, as mentioned, is proposing that the blue holes of Andros and Abaco become a UNESCO World Heritage Site.[28] Yet, in spite of these apparent synergies, there appear to be tensions between scientific researchers and the tourism industry.

More and more frequently, blue holes are on the radar of the more adventurous visitors to Abaco and Andros, and some drivers and tour operators know how to take people to find the holes. The main source of tension comes when visitors attempt to access valuable holes in improper ways. Researchers feel strongly that people shouldn't just hop in any hole on a whim. They need to be trained if they are to dive, and even then they need to be guided or else it's too dangerous. And they shouldn't disturb holes with known artifacts because they may be unwittingly destructive. But this apparent tension masks a deeper affinity between tourism and science concerning the holes.

In the spring of 2016 I visited Abaco once more to follow up on blue hole events. One morning I met up with the local research coordinator for the AMMC in her office in Marsh Harbour. When asked about the current relationship between research and tourism around blue holes, she explained that the AMMC can't take control of blue hole research alone due to lack of personnel, expertise, and resources, so they need to collaborate with many foreign scientific institutions and disciplines to learn about all aspects of the holes. They then hope to use the knowledge produced about the blue holes to promote them as unique tourist destinations to ensure their protection from development. The Ministry of Tourism is very interested in blue hole research as a feature to attract visitors to the sparsely visited Family Islands, and they have a great deal of influence when it comes to promoting blue hole protection to safeguard their attractions. But the ministry also lacks the resources and expertise to control the holes by themselves, she said.

In sum, blue holes are not accessible features for tourism without interdisciplinary science to narrate them and make them visible through practices of place. And blue holes are not safe from development or overexploitation without government-sanctioned protection enabled by their perceived value for the tourism industry. Thus, tourism and science meet within the notion of heritage as "science history." This is where tensions turn into creative collaborations because each industry is dependent on the other for the creation and perpetuation of valuable Anthropocene spatial formations.

Marcus, the co-owner of a local cave-diving company, embodies this collaboration.[29] He is a middle-aged American who relies on years of personal experience as a research diver with extensive knowledge of local caves and the latest cave-diving gear to legitimate his business venture. His small company offers guided experiences for visiting divers to inland and ocean caves, as well as training, photographic, and exploration dives. Marcus loves the caves of Abaco and has spent many years of his life there, exploring means to preserve the holes from future development. On that same 2016 visit to Abaco, this cave-diving entrepreneur took me to Dan's Blue Hole.

Dan's Hole was once hard to find, the narrow entrance hidden among heavy underbrush at the base of a small cliff in the pine forest. But now Dan's Hole is marked by a wooden platform and picnic tables built next to a small path culminating in steps down to the translucent surface of the water. Once again I found myself deep in a Bahamian pine forest, on the edge of a blue hole, waiting for a diver to surface. In this part of Abaco, an extensive cave system is revealed only in tiny surface openings like Dan's Hole, scattered throughout the pine forest. Marcus was diving to retrieve two sensors he had installed the day before, deep in the subterranean tunnels. In an attempt to map the system from the surface, two helpers from the National Museum used

FIGURE 15. Steps leading to Dan's Blue Hole, Abaco. Photo by author (2016).

detectors to locate the sensors in the forest, marking trees with red plastic ribbon in the places where they identified the signals percolating up through the limestone, thereby connecting an aboveground location with a location belowground and underwater.

This is all part of Marcus's efforts to eventually map the cave system below and above ground to interpret the space for visitors who do not dive or who are not experienced enough to cave dive. Marcus has a vision: there will be well-maintained trails running through the pine forest above ground that correspond to the twists and turns of the cave system below, with interpretive signs announcing the belowground location of an exciting cave feature, such as the caverns he has named the "Badlands" or "Fangorn Forest." These spaces will then become part of the larger narrative linking other holes on the island and the country to the artifacts and fossils collected and displayed at the soon-to-be-built Abaco Natural History Museum.

Blue Holes National Park was legally established in Andros in 2002, but Marcus's vision was recently enabled in Abaco by the government's legislation of a new mixed-use protected area in this location, South Abaco Blue Holes National Park. This park is the fulfillment of one of his dreams and a major goal for the AMMC. Marcus hopes that this park will eventually allow the AMMC to collect fees and tourism-related income to preserve the Abaco holes, and his business venture, in perpetuity.

These events show that interdisciplinary scientific research both redefines and rediscovers spatial formations, creating possibilities for the grounding of new national identities in new kinds of places. But beyond Bahamian nationalism, these identities and blue hole–based socioecological histories are now valuable in a global travel market that has interests in what it means to live in the Anthropocene. My experiences—in Andros helping a homegrown science to unfold and training young people to conduct and recognize ethnographic authority, and in Abaco observing different forms of hole documentation with interdisciplinary experts—merge to inform my awareness that one of the drivers of all this interest in blue holes will likely always be tourism.

Blue holes are examples wherein socioecological arguments about evolving planetary life underpin political understandings of the lives of Bahamian people. The stakes for blue hole research involve nothing more or less than the way Bahamians and visitors come to value their relations to their surroundings, to one another, and to other organisms. Blue hole research also points to the emergence of a form of ecobiological citizenship, wherein the foundation for collective cohesion moves beyond the consideration of merely human biology to include other forms of life.[30]

Nature and culture are elided—even exploded—in this arena in ways that are amenable to the diversification of The Bahamas in a global tourist market. These field exercises point to the development of blue hole–based scientific practices in which the research tools from each discipline combine to form an array through which blue holes become knowable and desirable. It is not just that blue hole nature becomes more cultural and therefore marketable but that holding the categories of nature and culture apart as distinct is no longer useful.[31] The marketable details the tourism industry searches for to diversify the Bahamian "tourism product" benefits from the universalizing globalism and simultaneous locality of place in the GCS narrative of the holes, but the social histories I helped produce are just as culturally specific as the natural histories I helped perform.

These notions of identity and travel markets are not unabashed celebrations of expanded possibilities. There are constraints at work within blue hole socioecologics too, something we realized while conducting interviews on Andros with Androsians who have spent their lives in blue hole country.[32] As the international GCS community and dive operations brought more attention and more people to the blue holes, these new visitors were characterized as "foreign tourists," which, in Andros and elsewhere in The Bahamas, has implications of whiteness and upper-class affluence. Some residents of Androsian settlements told us that they didn't expect to go near the holes once they became the provenance of tourism, protected by the AMMC. In Andros, where the population is predominantly rural, of modest means, and Afro-Caribbean, the holes were perceived as becoming the recreation grounds of the non-Bahamian and certainly the non-Androsian.

The irony is that the very people being studied to create a place-based social history and heritage for the holes—older, rural, black Bahamians with deep social memories—are the very people whose everyday relationship to the holes, fraught as it was, may be severed by the inclusion of these features in the tourism product of the country. And once severed through the authority of expertise within designated protected areas, that relationship can then be reimagined. In other words, long-time residents of blue hole country can become repositioned as those who lack sufficient knowledge about blue holes to maintain them for the benefit of science and tourism. Islanders must now learn from visiting experts about protecting a once-familiar geologic feature that has now been redesigned into global space. If the past thirty years of tourism in the country has been modeled on the tourist enclave, where tourist and Bahamian spaces are highly segregated, it is no wonder that some Androsians and observers like myself fear the same will be true of blue hole preserves and tours.[33]

Back in 2009 I had a discussion with an American microbiologist working at the Abaco site. She explained that her interest was geomicrobiology with a "deep time" focus.[34] In her opinion, the blue hole project was an effort that would be difficult for a strictly "nationalist science" alone because one needs a quorum of scientists to study these things, and each discipline's work cannot be conducted in isolation. She speculated that the pace of science moves too fast and that no one nation could keep up with all the new developments in research and theory. Collaboration, therefore, was good for science in that it could attempt to approach a bigger picture. For her, collaboration was the most fun thing she could possibly do, even if this project was not the most cutting-edge work out there for a microbiologist.

I contrast the potential exclusion and repositioning of rural islanders and my brief conversation about fieldwork fun with the microbiologist to reiterate the point that all these productions around blue holes stem from the authority of expertise. Expertise is required for the creation of valuable, marketable, and visitable space in the Anthropocene. Paleontology, biology, microbiology, geology, and so on are all forms of expertise that describe Bahamian ecology, but the interdisciplinarity of this GCS blue hole project, the collaborative "big picture narrative," creates something far more valuable in contemporary travel markets than any one disciplinary revelation alone. The point is not that blue hole science is cutting-edge—it might not be. The point is that this collaborative research is the new cutting edge for scientific practices that separate old notions of pristine nature from new socioecological spaces in the Anthropocene while at the same time potentially estranging island residents from a feature of supposed world heritage.

Conclusion

Anthropocene Anthropology

ANTHROPOGENESIS

This book is ultimately a story about an archipelago in transition. Through several loosely connected examples across a number of islands, I have shown how certain projects and processes are *producing* the Anthropocene as an idea with the capacity to remake the world. This happens as responses to perceptions of global anthropogenic change converge to reshape symbolic and material realities in specific locations. The Bahamas is the site where I first became aware of these processes at work, but at first these were subtle changes. When I started participating in global change science (GCS) research in The Bahamas Islands in the early 2000s, no one would have been able to point out many local examples of science and tourism working together to create new destinations. Researchers studying the efficacy of marine protected areas considered tourism as an afterthought, if they considered it at all. The connections between the two industries emerged for me over time as I put disparate examples together. Today, however, the situation has changed. The Anthropocene has arrived.

To document this, we need only to look at a brand new megaresort and casino, built from the bones of older hotels on the Cable Beach strip in New Providence. The gleaming, pale towers of this five-hotel conglomerate thrust straight up from the coastline, dwarfing all other structures on the island, visible for miles around. I have visited this property many times during construction and after opening, and I am confident that no one on this property has ever said the word "Anthropocene" out loud. Yet there is evidence of the Anthropocene *idea* all over this resort.

Across the street from the resort's enormous air-conditioned convention center and asphalt parking lots, there is a "reclaimed wetland" with an inviting boardwalk and interpretive signs explaining the names and characteristics of local mangrove species. The sprawling golf course uses recycled "gray water" to stay green. Each room has a sensor that will turn off lights and air when guest rooms are unoccupied to save energy. But the example I find to be the most compelling is still being built underwater, just offshore of the immense property. Giant boulders form protective structures in the ocean to prevent the erosion of replenished beach sand, and at the base of these structures, beneath the sea surface, there are several large spherical concrete balls. The balls are riddled with holes. These are artificial reefs, designed to attract colorful fish to their inviting niches. The megaresort management has its own in-house conservation unit, and that team of employees is working with local marine nongovernmental organizations to develop coral nurseries to aid in coral restoration in the country. In addition to the newly attracted fish, this team is planting coral fragments onto their artificial reef structures for the education and delight of resort guests, who will be able to snorkel over the reef balls on a "coral safari" once the site is ready. They market this as "smarter, greener luxury."

I have seen the submerged reef balls from high up in the towers. They look like vague gray blobs under water from that distant vantage point. Looking out over sunbaked tourists on the beach while listening to the buzz of passing jet skis in front of the complex, I have imagined what this experiment in marine manipulation might do for the brand reputation of the resort and the visitor experience. Over the years of construction, I have interviewed nongovernmental organization staff who advised the resort on the nursery project and coral ecology. One day soon I will snorkel the site myself, assuming that it will remain open access for more than just hotel guests.

This megaresort is a far cry from classic ecotourism. Thousands of rooms, nine swimming pools, and a full-sized golf course on a development that covers four hundred hectares will never be ecotourism. But, for me, the property's attempt to brand a certain kind of sustainability is more than a form of rampant green washing, although it is certainly green washing on many levels. This is an extreme extension of the Anthropocene socioecologics that I chart in this book, in which sharp distinctions between what is cultural and what is natural are no longer relevant to product developers in the tourism industry and the designers who use the predictions of GCS to inform their infrastructural decisions. Wetlands are in crisis around the world. Energy and water are wasted. Coral reefs are in decline on a global scale. Small islands are in danger. These are debates about anthropogenesis heard everywhere. The tourism industry has taken heed, in a fashion. Examples of so-called sustainable tourism like this are a trend that is currently evolving faster than we can document it.

By now it should be apparent that I have a deep love of The Bahamas. This is a dynamic chain of islands, a constantly changing travel imaginary, and the lifelong home of colleagues and friends I respect and consider family. My love of the place has forced me to pay close attention to the way these islands are designed and redesigned to reflect international trends in travel and the generation of knowledge about life on Earth. This perspective forms the basis for my version of Anthropocene anthropology.

ANTHROPOCENE ANTHROPOLOGY

It has been said that the "Anthropocene" is a gift for anthropology and an opportunity for the discipline to take center stage in political debates about global change.[1] We should remember, however, that the Anthropocene, as a term, is not yet mainstream. Even its validity as a scientific keyword remains up for debate. Many GCS scientists are unwilling to abandon the idea of the Holocene altogether, and so the proposed temporal category of the Anthropocene must compete with alternative geologic categories, such as the recently designated "late Holocene" or "Meghalayan."[2] In the critical social sciences, more alternative categories for the present have been proposed, including the "Capitalocene," referring to the pervasive effects of capitalism as a driver of global change, and my personal favorite, "Chthulucene," identifying the messy interdependencies among all earthly life forms that shape life as we know it.[3] Despite these alternatives, I still choose to borrow the term "Anthropocene," although I am not invested in legitimizing the category within the sciences.

Throughout this book I have referred to the "Anthropocene" because it is the term that many GCS scientists think best encompasses the conditions of change that are apparent everywhere and that many of us are all too aware of. It is a term I borrow to examine what such perceptions of anthropogenesis make possible. After endless arguments about what to do with the Holocene, it is clear that the Anthropocene *idea* effects material reality, regardless of the official scientific validity of a verified new epoch. While geologists feud over the stratigraphic record, we can observe people right now, out in the world attempting to live with global change. Therefore, the contested Anthropocene idea is more trope than theory, more a contextual framing for perceived problems than a definitive diagnosis of the present. The examples in this book don't prove the existence of fundamental truths about the Anthropocene, The Bahamas, tourism, or scientific research; rather, they are juxtaposed in the hope that they will provide insight into the production of life in the contemporary world.

To follow the events that produce our evolving understandings of life on Earth, I propose a research orientation that I call simply "Anthropocene anthropology."[4] Anthropocene anthropologists, and indeed any scholar interested in the critical study of anthropogenesis, should be responsible for examining the Anthropocene idea itself as an object of investigation. We should not take the term for granted or ignore its failings, but we should discover what the idea makes possible.

The practice of Anthropocene anthropology starts with multicited research for the examination of transnational processes in motion.[5] My understanding is that The Bahamas is shaped by a constellation of events, logics, politics, and projects that form and reform island spaces and places, and I cannot trace the effects of the Anthropocene idea here from only one vantage point. Therefore, my narrative has moved between multiple interrelated GCS examples on various islands of The Bahamas. This book is also a process of unpacking the Anthropocene idea as it manifests in different sites, the practice of grappling with an anthropological puzzle. Blue holes, lionfish, marine protected areas, experimental island settlements, carbon neutrality—all of these are puzzle pieces that together begin to delineate a Bahamian archipelago that is no longer so easily recognizable as The Bahamas of "Paradise and Plantation." When considered through the lens of the international projects that attempt to remake the islands in ways that respond to perceptions of anthropogenic change, the Anthropocene Islands come into greater relief.

As scholars of the present, Anthropocene anthropologists must consider the disparate pieces that compose our time: the movements of people and creatures; the forms of logic and reason; the complex histories; and the narratives, practices, exclusions, collectives, conflicts, and synergies that come together to shape the landscapes we encounter in our work. This research occurs on a number of levels. At the level of international research paradigms, I have engaged arguments within the scholarly literature and with Rita Colwell, the founder of biocomplexity research. At the level of everyday "technicians of general ideas," I have engaged with researchers and entrepreneurs on the ground in The Bahamas to observe how Anthropocene socioecologics are built and contested in the field.[6] And on the more elusive level of self-knowledge or subject formation, I have considered the influence of the Anthropocene idea in the creation of something like an Anthropocene subject—a more than merely human being, a being who lives in a self-consciously socioecological style.[7]

While we collect our puzzle pieces, we must not forget that the material world has power. Assemblages of technologies, products, places, humans, and nonhumans produce meaning and emotional conditions. Matter is creative.[8] Therefore, Anthropocene anthropologists must cultivate an ethic of care in our work that moves beyond reductive models of autonomous personhood. We should question our own ways of knowing so as not to obscure the power of the nonhuman, the physical, the material, and the things that make up our research sites, that in fact manifest *as* our research sites. Through the lens of an Anthropocene anthropology, The Bahamas becomes more than a place undergoing the effects of global change—the islands themselves, and their evolving destinations, are constitutive of a phenomenon.

Anthropocene anthropology is not grandiose—in fact it has also been recently said that "the problem with naming a new epoch is that no one cares."[9] Yet it is a practical orientation. And even though the Anthropocene is a borrowed term, it remains a useful umbrella concept within which we can now talk about complex situations of knowledge production in places where the stakes take particular shape around fieldwork, travel, anthropogenesis, and life. For Anthropocene anthropologists, the Anthropocene idea is an object of study and a space for collaborative participation. It forces us to think about the industries of science and tourism as cultural and ecological world-making processes, and it forces us to pragmatically engage with projects and practitioners as they reshape specific locales.

DESTINATION ANTHROPOCENE

Of course, the new Bahamian megaresort described in the first section of this conclusion is not sustainable by most measures. It uses a vast amount of energy, resources, and imported food, and it is too early to say whether it will even be economically viable, competing as it does with another megaresort nearby and the many other resorts in the region. But as a redesigned destination replacing the old Cable Beach hotels of the 1960s and 1970s, the resort now offers something more. Just like the lionfish as alternative cuisine, the Island Academy field station, the speculative carbon-neutral Caribbean brand, the blue hole diving adventures, and the expanding network of marine protected areas, the megaresort embodies the sense that the Anthropocene is here to stay and that the tourism industry must partner with GCS and adapt. The Bahamas itself is remade through all this adaptation.

After years of participant-observation tracking changes in these islands, I can now diagnose a condition that I call "destination Anthropocene." This condition is not unique to The Bahamas. There are many locations around the world undergoing processes of redesign and reimagination—some subtle, some extreme—as a response to perceptions of the Anthropocene. For anthropologists like me, arriving at destination Anthropocene means that we must attempt to understand how our research sites are tied to transformations of space, products and markets, science and scientists, and to our own shifting scholarly politics. Therefore, the condition of destination Anthropocene is best understood as a state of becoming. Recognizing this condition is a form of responsibility.

What does that responsibility look like? There are growing numbers of scientists who see the Anthropocene as a very real and powerful event that deserves our attention as residents of planet Earth. This book does not deny the legitimacy of that request to understand and respond to the effects of cumulative human practices across space and time. However, it should now be evident that the Anthropocene idea works across multiple sites and scales from the abstractions of grand theorists to the quotidian research practices of field scientists to the commercial schemes of entrepreneurs working within transnational markets. The idea animates a positive feedback cycle: research about global anthropogenic change reveals more anthropogenesis, inspiring more policy recommendations and more accessible narratives about change, inspiring more research in new areas seen as gaps in the previous research, inspiring more policy and more narratives about change that legitimate scientific and commercial enterprise. As the Anthropocene idea expands in this way, it falls into the hands of many people who respond to it in different ways.

As a situated observer in the midst of these events, I walk a line between appreciation of the creative capacities of the Anthropocene and concern for the way responses play out in the context of the postcolonial Caribbean. As a result, this text has been decidedly ambivalent, and questions about the ironies, tensions, and ethicality of the events I describe run throughout the chapters. What is enabled by the Anthropocene idea as a platform for thought and action, and what is disabled at the same time? What happens to notions of accountability within the new worlds of opportunity the idea opens up?[10] What does it mean to reproduce and capitalize on postcolonial small-island fragility?

In The Bahamas, authors such as Angelique Nixon write about alternative forms of tourism that may guide the nation away from the "Paradise and Plantation" model and toward a destination that can bring its colonial and racist past into the light and contend with historical and contemporary forms of prejudice. Public intellectuals such as Nicolette Bethel use social media to discuss the current limits of Bahamian political agency and envision revolutionary forms of democratic engagement. Artists and activists such as Tamika Galanis create visual works for Bahamian audiences that speak to the many injustices of globalization for Bahamian social and ecological worlds.[11] These Bahamians, and many more, are already actively reimagining their own conditions of possibility. In the hands of thinkers like these, the Anthropocene idea will become more than an opportunity for the expansion of the tourism industry and scientific careers. These scholars have already created works of Afro-Caribbean futurism, made plans for locally focused small-island sustainable settlement life, and built human-coral hybrid sculptures that are not confined to museums or galleries. Their Anthropocene is distinct from the more conventional notions of the GCS sciences and the mainstream tourism industry and is best described in their own terms.

As for me, I want to resist the urge to create capital from the degradation of the world. As scholars who care about anthropogenesis, we should make sure our socioecologics do not reproduce colonial paradigms of scientific expertise that turn archipelagoes into vulnerable "small islands" and islanders into uninformed "locals." We must exercise caution in the face of the ever-expanding applied sciences.[12] GCS scientists create new understandings of planetary processes that lead toward positive changes in the way we live, work, and produce, but they can also perpetuate ahistorical and apolitical research situations. The chapters of this book have attempted to foster a sense of hesitation in the face of strident calls for prescribed action in the form of more protected areas, more resource regulation, more socioecological modeling, more international field research, more sustainable development projects, and more tourism as the economic engine that will bind all these schemes together. Such schemes hold us hostage, stifling actual thought, when in fact the capacity to modify our assumptions as we go can be our greatest strength. Therefore, the examples presented in this book are attempts to open up GCS and the evolving tourism industry to scrutiny from new angles as a form of decolonizing Caribbean scholarship.

Events in The Bahamas show that the Anthropocene idea should be a vital Caribbean question. Historically contingent regional questions of race, class, subjugation, systematic exploitation, and capital accumulation are now expressed through the material and symbolic politics of global environmental change. Evolving socioecologics are as critical for understanding contemporary Bahamian and Caribbean life as the effects of the plantation system, colonial rule, and postcolonial independence. All too often in the region, GCS research practices exacerbate social divisions that preexist or are even created by the studies in question. And in many cases small-island places construed as isolated, fragile, and vulnerable are also continually brought into international trade and knowledge networks, subject to the exclusions and prejudices that travel within them. It is my hope that this book about The Bahamas will inspire other scholars in other transforming places to conduct their own investigations of the Anthropocene idea, and together we might begin to delineate "emerging habitats in which a new mode of humanity might be nurtured into life."[13] There may be no other way to arrive at destination Anthropocene.

Notes

INTRODUCTION

1. H. Bell (1943).

2. See the anthropologist M. Estellie Smith (1977) and the Bahamian scholars Ian Strachan (2002) and Angelique Nixon (2015).

3. Kincaid (1988).

4. D. Campbell (1978a).

5. For example, see McNeill and Engelke (2014), Scranton (2015), Kress and Stine (2017).

6. Luhmann (1993).

7. See K. Thompson (2006) on palm trees.

8. Rabinow (2003).

9. The Anthropocene is another attempt that is similar to ideas such as "modernity" or the "end of history."

10. Relatively few sociocultural anthropologists have explicitly taken up the idea of the Anthropocene as an object of inquiry, although there is a large breadth of anthropological work centered on environmental change and climate from the early climate and culture studies of cultural ecologists (Steward 1955) to the current anthropology of climate change (Crate and Nuttall 2009; Crate 2011). Climate change itself has become a form of social fact, mobilizing labor, capital, and intellectual efforts into assemblages of technoscientific governance that bridge and create scales (Dalmedico and Guillemot 2009). Many anthropologists help to shore up the discourse and framings of the social fact of climate change with work that enriches internationally circulating ideas such as resilience, adaptation, vulnerability, and sustainability with local specificity (Nelson, West, and Finan 2009; Crate and Nuttall 2009; Galvin 2009; Kelman and West 2009). Other anthropologists investigate the international production of climate change as a social fact to understand its social and cultural implications (Barnes

et al. 2013; Gutierrez 2007; Fiske 2009; Lahsen 2005, 2010; A. Moore 2010; Tsing 2005).

11. Crutzen and Stoermer (2000).

12. Paul Crutzen and Eugene Stoermer wrote, "considering (the) . . . major and still growing impacts of human activities on earth and atmosphere, and at all, including global, scales, it seems to us more than appropriate to emphasize the central role of mankind in geology and ecology by proposing to use the term 'anthropocene' for the current geological epoch" (2000, 17).

13. While the Anthropocene is becoming an increasingly popular attempt to label the effects of the human species, climate change is still arguably the most well-known subset of anthropogenic change and is often used as a stand in for planetary anthropogenesis in general (Zalasiewicz 2010). The Anthropocene idea, however, has some important conceptual distinctions from climate change. As mentioned earlier, there are many forms of anthropogenic change encapsulated by the Anthropocene idea, with climate change being but one form. Further, the Anthropocene idea is explicitly tied to anthropogenesis in an unambiguous way (though the form of that anthropogenesis and the understanding of responsibilities are anything but clear). Climate change is used as an iconic referent for anthropogenic planetary change, but the idea of a changing planetary climate does not automatically necessitate anthropogenesis, and this has led to a great deal of controversy about the anthropogenic or natural origins of warming global temperatures (Farnsworth and Lichter 2011). The Anthropocene idea is an attempt to dispel that controversy once and for all.

14. See Crutzen (2002). Despite the growing consensus since 2000, there is disagreement about the origins of the Anthropocene, with some scholars pointing toward such distant events as the advent of agriculture or even the harnessing of fire as the key moment in human evolution that led toward large-scale anthropogenesis in the earth system (Gowdy and Krall 2013). Crutzen (2002) locates the origins of the Anthropocene in the late eighteenth century, coinciding with the rise of industrialization (Crutzen and Stoermer 2000; Crutzen and Steffen 2003). Since then, he and others contend, the rate of human-driven change and variability in the earth system has only increased, creating a "golden spike" of human activity in the geologic record, especially post–World War II (Waters and Zalasiewicz 2013). The fear inspired by this diagnosis is that rapid and unpredictable change will continue to increase into the foreseeable future unless humanity can transition to more sustainable endeavors (Crutzen and Steffen 2003; Furnass 2012).

15. There have been a number of recent conferences and lecture series convened in the name of the Anthropocene (at the University of California, Santa Barbara, Santa Cruz, and Berkeley; at Maquarie University and the Australian National University; at the University of Kent and the University of East Anglia; and at Stockholm University, to mention only a few), along with several academic journals *(Elementa: Science of the Anthropocene, Anthropocene, Anthropocene Review,* and *Anthropocene Journal).*

16. Waters and Zalasiewicz (2013).

17. See Kolbert (2014) on the sixth mass-extinction event.

18. Kloor (2013).

19. Caro et al. (2011).
20. Lorimer (2012).
21. See Robbins and Moore (2013). Some scholars read the Anthropocene as symbolic of the moral stance toward an anthropogenic crisis oriented around the practical limits and boundaries of human activities and economies (Rockstrom 2009). New forms of social media have been created based around the idea that the stakes of the Anthropocene concern nothing less than the morality of contemporary environmentalism and the possibility for real political recognition of planetary uncertainty (Osborne, Traer, and Chang 2013). The notion has been hailed as a revitalization of environmentalism, injecting new vigor into a passé movement, rearticulating science and society relations, forming new technoscientific assemblages, and promoting emergent disciplinary methodologies and research questions (Seidl et al. 2013).
22. Slaughter (2012).
23. The designation of the Anthropocene as the present earth epoch by the International Commission on Stratigraphy is pending at the time of this writing. The Anthropocene Working Group has stated that "the 'Anthropocene' is currently being considered . . . as a potential geological epoch, i.e. at the same hierarchical level as the Pleistocene and Holocene epochs, with the implication that it is within the Quaternary period, but that the Holocene has terminated. It might, alternatively, also be considered at a lower (Age) hierarchical level; that would imply it is a subdivision of the ongoing Holocene Epoch" (Subcommission on Quaternary Stratigraphy 2015, 1).
24. As a more "specific" concept and subset of Anthropocene events, climate change has led to distinct propositions, from carbon trading to geoengineering (Galaz 2012; Dalsgaard 2013), although to date these responses have not yet led to changes at a scale that might significantly halt the rate of planetary warming. Like climate change, however, the Anthropocene idea vacillates between apolitical claims about human-driven change—for example a news feature in *Nature* states, "through mining activities alone, humans move more sediment than all the world's rivers combined. *Homo sapiens* has also warmed the planet, raised sea levels, eroded the ozone layer and acidified the oceans" (Monastersky 2015, para. four), with no specificity about which humans might be more responsible than others for certain forms of change—and apocalyptic claims of impeding planetary collapse and the end of life as we know it. A commentator in *Forbes* posits, "in the end, [the Anthropocene] debate might be shear hubris since when we are gone, future geologists, of whatever species, will decide for themselves where they want to place the beginning of this particular catastrophe" (Conca 2014, 1). And so, like climate change, while scientific knowledge production animates the idea, its meaning will not be dictated by scientists alone but by the experiences and engagements of multiple publics around the world. In sum, the Anthropocene idea is both stronger and more vague than the concept of climate change and therefore open to multiple interpretations, uses, and tensions (Swanson, Bubandt, and Tsing 2015).
25. For example, see Chakrabarty (2009, 2012, 2013); Holm et al. (2013); Latour (2013); and Lorimer (2012). The postcolonial historian Dipesh Chakrabarty (2009) makes the point that the very idea of the Anthropocene is a challenge

to critical thought in that it influences the way the human is conceived as the subject of scholarship in the humanities. He posits that the twentieth-century scholarly theme of globalization has been forever altered by the notion of anthropogenic planetary change (climate change being his primary focus) because the idea of humans as a geologic force cannot be explained in terms of critiques of capitalism or enumerations of forms of social difference alone. The human itself must now be acknowledged as some kind of entity, some kind of "we," referring to our more-than-human capacities as a "species," for lack of a better term.

Along these lines the science studies scholar Bruno Latour has argued that thinking about the Anthropocene means rewriting the typical scripts of human progress that have unreflectively justified so many grand economic ventures. He writes, "what makes the Anthropocene a clearly detectable golden spike way beyond the boundary of stratigraphy is that it is the most decisive philosophical, religious, anthropological and, as we shall see, political concept yet produced as an alternative to the very notions of 'Modern' and 'modernity'" (2013, 77). The geographer Jamie Lorimer (2012) explains that these revelations require new research paradigms to examine integrations of the human and nonhuman world across multiple scenarios. His point is that researchers should no longer assume that the human or nonhuman elements of these scenarios exist a priori of their interrelation (see Holm et al. 2013).

26. Chakrabarty (2012, 2013).

27. Moore, Kosek, and Pandian (2003).

28. The way current problems are institutionally framed and discursively tied to pasts, presents, and futures demands critical attention, because this is how Caribbean people and nonhuman life forms become conscripted by powerful political forces (Scott 2004).

29. My orientation owes a debt to a number of scholars who are bringing the Anthropocene into relief for critical analysis. For example, S. Eben Kirksey and Stefan Helmreich (2010) ask that anthropologists take seriously the task of writing culture in the Anthropocene (following Clifford and Marcus 1986; see also Irvine and Gorji 2013) to track the poetics and politics embedded within evolving networks of global change. Nathan Sayre points out that "the key points to draw from the Anthropocene have less to do with when it began than how it affects the underlying assumptions that scientists make about understanding the world" and how these assumptions affect possibilities for justice and policy (2012, 63). And a number of anthropologists have made the case that forms of governance are evolving in an Anthropocene context, leading to the creation of new global socioecological assemblages (Ogden et al. 2013); systems of thought that enable novel forms of governmental intervention (Lovbrand, Stripple and Wiman 2009); and authoritative truth claims about the future (Howe 2014).

30. A. Moore (2016); Latour (2013). See also Rabinow, who asks, if the forms of reason that ground our understanding of life are shifting, then "what logos is appropriate for anthropos?" (2003, 14).

31. International Geosphere-Biosphere Programme (2010).

32. Personally, there is very little these researchers have in common, although they share the biases of the academy and are more likely than not to be heterosexual, white, and male—with notable exceptions.

33. National Science Foundation (2015).

34. Strang (2009); Barry and Born (2013).

35. Socioecology originated in the mid-twentieth-century behavioral sciences as the study of social organization between species and between species and their environments (Rubenstein and Wrangham 1987).

36. Socioecologics run between the poles of what is called complex adaptive systems theory and socioecological systems theory (see Holling 2001; Anderies, Janssen, and Ostrom 2004; and Berkes 2003). This style of thought can take many forms, but one example is biocomplexity research, also called coupled natural and human systems research.

37. Costanza, Segura Bonilla, and Martínez Alier (1996); Martin and Schluter (2015).

38. The Anthropocene is challenging on multiple levels. Life conceptualized within the socioecologics of Anthropocene GCS science is more than just the "life itself" understood as a molecular or genomic human bodily process (Rose 2006). Human life is now scientifically understood simultaneously as a postgenomic milieu in a biomedical context (Richardson and Stevens 2015) and as part of a holistic socioecological system known through specific GCS forms of knowledge production designed around managing environmental anthropogenesis. For GCS scientists the idea challenges the business-as-usual economic practices of contemporary industrial societies by making the planetary limits to human actions tangible as threats to existence as we know it (Wackernagel et al. 2002). For social scholars the idea challenges central modern themes of natural and cultural dualism and separation by conceiving of human activity as a force of nature akin to existing global processes, eliding long-held human and nonhuman distinctions (Singleton 2012).

39. This line of reasoning can also be explained as biopower versus environmentality. The utility of environmentality (Agrawal 2005) as a useful concept for analyzing the formation of subjectivity within intercultural environmental management has been challenged by Cepek (2011), following ongoing critiques of the utility of governmentality (D. Moore 2000; Kipnis 2008).

40. The medical anthropologist Valerie Olson (2010), thinking from the field of space biomedicine, has chosen the somewhat unsurprising term *ecobiopolitics* as a concept capable of acknowledging techniques of knowledge production and governance that attempt to redesign the interface, or milieu, between environment and human processes for the optimization of the milieu itself. As a conceptual category, ecobiopolitics helps explain practices that manage the idea of "habitable space" and elide human life and ecological understandings.

41. Concepts such as ecobiopolitics and socioecologics speak to the Anthropocene by reorienting our attention to the GCS institutions that seek to manage this broadened notion of life, moving our critical capacities beyond the gene and beyond the human to think about emergent forms of connection. The Anthropocene idea reveals most scholarship on biopolitics as too human centered to speak to GCS worlds. In other words, a great deal of notable work (Rabinow 1999; Rose 2006; Campbell and Sitze 2013) has intentionally been so heavily engaged in the implications of bioscience as it relates to human bodies that it cannot go far in explaining transformations within conceptions of life

understood in a planetary and systemic frame. Socioecologic and ecobiopolitics can also move concerns with biopolitics more squarely into the material world of organisms, landscapes, ecologies, research objects and technologies, enabling critical attention to the every day "stuff" that participates in and is produced by Anthropocene politics (Braun and Whatmore 2010). We must not lose sight of the fact that these conversations are more than mere abstractions.

The human-centered "bios" of the concepts of biopolitics and biopower and the nature-centered "eco" of ecopolitics themselves have been destabilized in the Anthropocene, replaced with a more totalizing notion of life that is as yet amorphous and in formation. Conceptions of life are now an open question of moving beyond the human to create new forms of citizenship, belonging, and alliance (see Latour 2004; Haraway 2008; Helmreich 2009; and Kirksey 2014 as just some of the scholars and works that address this question directly). But rather than the human disappearing entirely, the Anthropocene describes the development of socioecologics that reshape what it means to be human today, and it calls for an attention to the ways that science can formulate nature (including and especially human nature) as environmental, ecological, geologic, and planetary *in conjunction with* the bodily, the genomic, and the molecular. The notion that human citizenship and sociality has thus become a scientifically mediated question of biology, community, population, lineage, race, and species, and not just of membership within a nationally defined social body (as explicated by Rabinow [1996] and Rose and Novas [2005] through the terms *biosociality* and *biological citizenship*) is so far only the beginning of the story (see also Mallard and Paradeise 2009).

42. See Darwin (1839, 1859).

43. The study of islands and island regions in anthropology is not new. Foundational anthropological work in the Pacific considered islands as sites for the study of "dying cultures" and isolated peoples. Caribbean islands have historically been used as sites in which to identify and study the social process of mixing (acculturation and creolization are only two of several famous examples). And now we can see that islands, imagined as particular kinds of natural and social spaces, have been imbued by many scholars with a mystique and significance that continues to animate contemporary problems, especially the multiple crises of the Anthropocene. On the subject of the planet as island, see Lovelock and Margulis (1974) and Meadows et al. (1972) but also Olson and Messeri (2015).

44. Pugh, Chandler, and Stratford (2015), para. 2.

45. Lemov (2005), describing Bacon (1624).

46. L. Young (1999, 250).

47. Kirch (1997, 30).

48. Lemov (2005, 22–23). Islandness as a style of looking can be seen in the work of many geologists and ecologists, especially population biologists (for example, L. Young [1999] and the classic Wilson and MacArthur [1967]), who see islands as important for both human and other life because of their isolation and accessible scale. See also Deloughrey (2013).

49. See Bahn and Flenley (1992). Islandness as a style of looking has been profoundly influential, especially for the development of anthropology itself. The

study of islands was especially central in terms of the underexplored relation between the island model of Darwinian biogeography and British anthropology (Kuklick 1996). For nineteenth-century European natural historians, islands were considered key places to study natural selection and rare species, and they then became so for the study of "rare" human cultures. Island humans were then and are now again considered to be just as vulnerable as plants and animals, and islands became valuable for social science as isolated "virtual laboratories for natural experiments in social processes" (617). This is the Pacific island tradition of which Malinowski (1922) and Radcliffe-Brown (1922) are a part.

50. Thomas Malm (2006) critiques the general characterization of the earth as an island in space, a characterization highlighted in the discourse of the Anthropocene (take, for example, the *Earth Island Journal*). He notes that islands are a popular analogy for isolation (the laboratory move), but he questions this analogy, stating that no island is truly isolated—that no island is actually "an island." For a dramatic example of the material integration of an entire island into capitalist networks, see Teaiwa (2014).

51. There is already a healthy body of anthropological literature on the subject of small islands and climate change (Lazrus 2012).

52. Mintz (1983); Sheller (2003).

53. Boruff and Cutter (2007); United Nations (2005a).

54. Nature Conservancy (2016); World Wildlife Fund (2016).

55. When foreign scientists say this I find that they are usually implicitly referring to Jamaica, Haiti, and the Dominican Republic—nations assumed to have overexploited their natural resources and to have practiced "fishing down the food chain" to a great extent. This sentiment unfortunately aligns with the popular sense of Bahamian superiority in the region, owing to the relative stability of the economy over the majority of the twentieth century and the widespread prejudice held by some Bahamians against Haitian and Jamaican immigrants to the country.

The ways in which The Bahamas does and does not exercise its "Caribbeanness" are complex, and one should not assume that this is a completely "Caribbean place," even if these islands share a Caribbean history. There is a politics to being (and not being) Caribbean. Caribbeanness is a complex attribution with which The Bahamian state and Bahamian people continually grapple. Historically, The Bahamas has been subject to the same wide-ranging and influential events as the rest of the Caribbean region; most notably the transatlantic slave trade, European colonialism, and the twentieth-century independence movement. Yet The Bahamas has been excluded from many collections of social science on the Caribbean and is usually not listed as a Caribbean country when academic scholars discuss the countries of the region, though it is often categorized as part of the Caribbean region when it comes to international state politics. For example, the United States' Central Intelligence Bureau lists it as a Caribbean nation. This confusion results from more than the fact that The Bahama Islands are not in the Caribbean Sea (they are situated just north of Cuba on the North American Plate). It has been written that Bahamians do not consider themselves Caribbean and especially not West Indian because their social affinities and trade ties lie more with the United States than with the other islands (Bethel 2008). It has also been written that The Bahamas, owing to its long history of success with tourism, is too wealthy to be classed with the rest of the Caribbean, even the Caribbean of former British colonies. The Bahamian government has reservations about its membership with the Caribbean economic community and its subsequent inclusion in the Caribbean free market.

56. How should we characterize such island stories? In terms of a framework for thinking about the Caribbean region and forms of alliance and association articulated there, I am intrigued by Antonio Benitez-Rojo's (1996) trope of the repeating island. Within the repeating island the Caribbean is held together as a region by the historical flows of the colonial plantation, making the region's experience with subjugation and the world system explicit. Instead of pointing to islander connectivity and resilience in the face of change, the repeating island framework points toward a much more ambivalent sense of connectivity. The repeating island is a notion meant to evoke the impossibility of characterizing the region with an attention to one site or the comparison of diverse sites. Rather, the sense of the repeating island—as infinitely repeating and inescapably complex—is the feeling one gets when one tries to pin down the culture of the region, and the feeling is far more real than any regional "essence."

From another context the anthropologist Bill Maurer (1997) has argued that proponents of globalization theory often looked to the Caribbean as a symbol of global processes. The Caribbean region, with its transnational flows of people and modes of accumulation based on movement, seems to embody and stabilize the very idea of globalization, and he referred to this popular characterization in scholarly literature as the "Caribbeanization of the world"—the way in which the Caribbean has been used as an example to reify fluid notions of identity, belonging, and forms of exchange.

57. Kelman and West (2009); Rudiak-Gould (2010, 2013).

58. A. Moore (2015a, 2015c, 2016).

59. There are a number of ecobiopolitical articulations of spatial alliance underway in the name of planetary change that are not only promoted by transnational GCS but exist alongside or contest small-island vulnerability indexes and categories. The Alliance of Small Island States (the body that coordinates island nations designated as Small Island Developing States) is one political organization asserting islander agency in the global debates about planetary change. Another voice comes from the Sea of Islands framework for conceptualizing Pacific island peoples promoted by Epeli Hau'ofa (2008) and applied within the context of climate change by a number of scholars (Lazrus 2012; Barnett 2002; Farbotko 2010), in which, again, islanders are known not by their isolation but by their networked connections to other peoples and places through "the flows of island people, materials, and goods that have always made island living possible" (Lazrus 2012, 289).

60. Understanding the Anthropocene Islands means examining key forms of "writing islands in the Anthropocene" (Graziadei and Riquet 2015). In a similar vein Annemarie Mol and John Law, when talking about human bodies, state that "in practice, if a body hangs together, this is not because its coherence precedes the knowledge generated about it but because the various coordina-

tion strategies involved succeed in reassembling multiple versions of reality" (2002, 10).

61. Hare (2015).

62. Pilgrim (2014); Hanna and McIver (2014).

63. For example, the ocean explorer Jean-Michel Cousteau posted in a blog to document his extensive trips to small islands in the Pacific, "like the canary in the coalmine, thousands of drowning islands in the Pacific are telling us that something dangerous is happening. As ocean levels continue to rise higher onto their low-lying lands, millions of people are facing a reality that threatens their homes, their families, their economies and their entire way of life" (Cousteau and Mandoske 2014, para. 1).

64. Saunders and Craton (1992, 1998).

65. World Travel and Tourism Council (2017).

66. Jefferson and Lickorish (1991, 7).

67. While the promotion of The Bahamas as a mainstream U.S. tourist destination is not often directly tied to the creation of these parks, former tourism official Angela Cleare, whom I use as a stand-in for the ministry in general, makes the connection, stating, "the success of tourism is linked to a healthy environment" (2007, 122). Cleare cites as centrally important the fact that The Bahamas is signatory to Agenda 21, the program of action stemming from the 1992 United Nations Conference on Environment and Development, as well as the fact that the First Conference of the Parties of the Convention on Biological Diversity was hosted by The Bahamas in 1994. All this activity made late twentieth-century environmentalism a large part of the play of arrivals and visitation in the country and prepared the ground for the rise of Anthropocene discourse and practice.

68. Cleare (2007, 34).

69. Lahsen (2005); Helmreich (2009).

70. Robert Kohler's "practices of place" (2002) describes the experimental manipulation of place through ecological fieldwork when field research practices involve the reconstruction of the histories of living processes in specific places and observation and comparison between places. Such practices invent tools to read and know nature. In terms of contemporary socioecologics, people and their activities are part of the process of site selection now that ecologists want to observe the effects of people on the environment. In this way the crises of the Anthropocene become research opportunities, and people living in a place become a part of the processes of knowing place for field scientists.

As a historian of science, Kohler considers the labor that constitutes the field when he describes ecological fieldwork developed in the twentieth century through practices combining field- and labwork in the form of field stations and specific important places. Place becomes a tool of field scientists in that they work on place to create place; place itself becomes an object of their work. As Cori Hayden (2003) has also described, field scientists turn the uniqueness of natural places into an advantage, and this practice becomes even more obvious in the Bahamas, where research and research-related activities are all about the specificities of island and marine places.

71. See Raffles (2002) on materialities of the obvious.

72. UNWTO (2008); Blue and Green Tomorrow Communications (2013).

73. A. Moore (2010).

74. West (2008); Lemelin et al. (2010); Norum (2013).

75. Strachan (2002).

76. See Davidov (2012) and Ogden et al. (2013) on intersections of science and capital within global assemblages (Ong and Collier 2005) for the Anthropocene. Further, scientific research conducted in the name of the Anthropocene cannot focus strictly on human or nonhuman realms of knowledge production but must instead integrate these realms in ways deemed "policy appropriate" and open for intervention by governance and commercial enterprise. Resource-based economies are internalized in the process of environmental research (the study of disappearing marine species becomes the study of fisheries; the objects of climate research become features for green tourism) and older, politicized forms of science and capital integration, such as the biodiversity assemblage (West 2006; Hayden 2003), become reformed as biocomplexity, matching older solutions (conservation through protected areas and property rights) with emergent problems (unsustainable and uncertain human and environmental systems).

77. Agard-Jones (2013).

1. BUILDING BIOCOMPLEXITY

1. I have chosen not to name this project directly to discuss it without implicating its creators and participants.

2. All the names in this chapter have been changed, except for those belonging to public figures speaking publicly.

3. *Oxford Dictionaries Online,* s.v. "biocomplexity," accessed January 22, 2019, https://en.oxforddictionaries.com/definition/biocomplexity.

4. *Oxford Dictionaries Online,* s.v. "biodiversity," accessed January 22, 2019, https://en.oxforddictionaries.com/definition/biodiversity. See also Takacs (1996).

5. See Hayden (2003).

6. Kohler (2002).

7. Scoones (1999).

8. For socioeconomic formats that aggregate individual responses to preset questions about the costs and benefits of daily life, see United Nations (2005b).

9. See the introduction to this volume and Olson (2010).

10. Example questions from the socioeconomic survey are paraphrased here to protect the identity and intellectual property of the survey creators.

11. Rabinow (1996).

12. Raffles (2002); Haraway (1989).

13. Part of this process of re-creation includes the way the study of "bios" (Rabinow 1999) has been transformed by Anthropocene GCS in recent years within notions like biocomplexity.

14. Colwell (1998).

15. Colwell (1998, 786).

16. Michener et al. (2001, 1018).

17. Fujimura (2011).

18. The primary commercial fish species are the Nassau grouper, Caribbean spiny lobster, and queen conch, together described as the Bahamian "holy trinity."

19. This observation comes from my June 2008 interview in Washington, DC, with Dr. Michael Mascia, then head of social science at the World Wildlife Fund.

20. To protect the anonymity of certain individuals, projects, and organizations, I have omitted citations for their quotations from reports, documents, institutional idioms, mission and vision statements, speeches, and marketing copy. Anonymity is required to maintain the identity of those who work on sensitive projects and who should not be critiqued as specific actors, only as examples of more generalizable processes.

21. BenDor et al. (2014); Rounsevell, Robinson, and Murray-Rust (2012); Smajgl et al. (2011).

22. The term *human* is also popular for its extreme generality. For example, see Latour (2005).

23. See Strathern (2005); and Nowotny (2001). These works describe the introduction of performed social accountability into the academic sciences, institutional arenas previously seen as outside of the boundaries of social accountability.

24. See Paley (2001).

25. There is now a legally mandated annual closed season for Nassau grouper in The Bahamas from December to March.

26. Queen conch are currently listed as threatened under the Convention on International Trade in Endangered Species; see "*Strombus gigas*" (n.d.).

27. Hayden (2003).

28. Hayden (2003).

29. Paley (2001).

30. Strathern (2005).

31. I think we must think of biocomplexity—and the refashioning of lives, livelihoods, methodologies, and subjectivities that the concept evokes—as a contemporary Caribbean problem. Indeed, the Caribbean itself is evocative of certain complexities, systems, and socialities. In this case it is appropriate to revisit Antonio Benitez-Rojo's *Repeating Island* and his proposal to investigate "Caribbeanness" and the Caribbean as a cultural "meta-archipelago," having no boundaries and no center, only a constant flow "outward past the limits of its own sea" (1996, 4). This region is a great machine built by the European plantation economy. The Caribbean machine repeats itself continually, reproducing the plantation society over and over. This cultural meta-archipelago is thus the paradigmatic site of contact and diaspora, having always been involved in processes of continual global dynamism.

The GCS redefinition of the human and environmental milieu of The Bahamas as a model biocomplex system—and therefore as calculable and potentially predictable—is a contemporary repeating island. The project's designers share Benitez-Rojo's concern for the discovery of some sort of rhythmic pattern within seemingly chaotic events. For Benitez-Rojo these events are sociocultural, while for GCS researchers self-consciously working in an Anthropocene context, they are socioecological. The project created its own Caribbean "meta-archipelago" out of the Bahama Islands, a repeating field laboratory for the study and management of anthropogenesis, by linking proposed MPAs in an archipelagic system of Marine Reserve Network sites to comparable sites of human island settlement, to repeating occupational categories, and to populations of mobile marine species in a program of research intended to project its findings into the future. The project hoped to create a replicable and predictable MPA-based management package out of preset patterns of life and livelihood.

The Caribbean and its repeating islands have long been the site of anthropological studies that have sought to use the region as a symbol for larger global processes, such as globalization. Bill Maurer (1997) argued that the Caribbean had become an example and symbol of the "postmodern" condition, of the continual revision and rearticulation of culture within a globalized world. He cited Benitez-Rojo on this subject and also James Clifford (1988) on the notion of the inventive "Caribbean experience" as contrastive to "monolithic" notions of culture held by such theorists as Claude Lévi-Strauss. This reasoning is problematic for Maurer, however, because it is a mythology that unwittingly promotes other forms of exclusion through the invocation of biological hybridization and recombination implied by the notion of "Caribbeanization."

32. Nature Conservancy (2017).

33. The socioecologics of biocomplexity transform the notion of biopolitics and the related notion of biosociality, shifting the conceptual biopolitical focus from the formal knowledge politics of the human to emergent conceptual linkages between the life processes of living organisms and the living practices of human societies (see Michel Foucault, as described in Rabinow and Rose [1994] and Rabinow [1996]). Human health is now an ecological question of planetary systems and complex ecosystemic environments as much as it is a biomedical, bodily concern. These global systems are arenas in which the nonhuman plays a central role. The *bio* of biocomplexity therefore points to a series of spatial and temporal relationships between chemical materials, organisms, socialities, and systems—in other words, life becomes a biogeochemical and social totality, a holistic cosmology, a complex milieu. And within this emergent ecobiopolitical view, instrumental political connections between the realms of economics, science, and culture are designed anew (Olson 2010).

34. As Anna Tsing (2005, 105) writes, "The common assumption is that everything can be quantified and located as an element of a system of feedback and flow."

35. For examples, see Committee (2001); McClanahan, Davies, and Mania (2005); and Pomeroy, Parks, and Watson (2004).

36. "Bahamas" (2015).

37. West and Carrier (2004); Carrier and Macleod (2005); Sommer and Carrier (2010).

38. "Conservation discourse" is also known as "conservation as development" rhetoric (see West 2006).

39. Hall-Campbell (2007, 56).

40. Mascia, interview.

41. Braun and Whatmore (2010).

2. THE EDUCATIONAL ISLANDS

1. Very few Bahamians trace their lineage back to indigenous groups.

2. Saunders and Craton (1992).

3. All names have been changed.

4. The University of The Bahamas was formerly the College of The Bahamas until 2016.

5. The public schools in The Bahamas and on Eleuthera are staffed primarily by Bahamian women trained in education, and some of the more remote islands have trouble obtaining and retaining teaching and support staff for their schools. During the colonial era many schools were staffed by white British teachers, many of whom traveled to the remote islands of The Bahamas for that purpose. A public school education consists of basic reading, writing, mathematics, and biology, though some schools have more of an emphasis on specialized disciplines and high school courses provide more advanced exposure to chemistry, foreign language, physics, and so forth. There have also been successful efforts made by the Bahamas Reef Environment Educational Foundation and the Bahamas National Trust to include curricula on the Bahamian marine and terrestrial environment and ecology in standard public education, and the foundation runs an annual teacher-training field course for Bahamian teachers to this end.

6. These centers undergo their own cycles of boom and bust as a result of shifting funding and staffing patterns.

7. Much of the archaeology is based in the work of William Keegan (1992) on Lucayan sites.

8. I was told that the academy tries to buy groceries locally, but that there is usually not enough to be found on the island to feed a hundred people consistently. The guide told me the school spends $20,000 to 30,000 dollars per month on food alone.

9. The price of academy biodiesel fluctuates with the price of oil because their production process uses methanol, which is tied to the price of oil.

10. Cotton Bay has since failed.

11. Interviewing the students would have violated the terms of my research permit and the norms of human-subjects research, wherein it is generally unacceptable to interview anyone under eighteen without special dispensation. I encountered the Island Academy accidentally during the serendipity of fieldwork, and so I did not get special dispensation to talk to minors from my university's research review board during that visit.

12. Island Academy, mission statement, 2010.

13. Island Academy, vision statement, 2010.

14. This example represents what Jan Laarman and colleagues might have referred to as a blending of "soft" and "hard" nature tourism. Soft nature is the realm of ecotourism, they say, with some education mixed with mild adventure. Hard nature is the realm of professional researchers, professors, and students who travel to study "serious science," usually tropical biology. Laarman and others also refer to the science tourist as "Caucasian, male, and highly educated" (1989, 213). The advantages of the science tourism of those interested in hard nature, the authors say, is that, as a generally "wholesome activity," it

tends to be complementary with natural resource conservation, to tolerate limited physical infrastructure, and to provide educational benefits to the host country in terms of environmental education.

15. See Paige West (2008), who initiated the discussion. There are few observations in the loosely defined anthropology of tourism that mention scientists as tourists (but see Laarman et al. [1989] and West [2006]).

16. Angelique Nixon (2015) also argues that almost anyone who travels via the infrastructure and forms of mobility enabled by the tourism industry is a participant in that industry, even if they might classify themselves as a student or scholar.

17. Valene Smith (1989, 1) and Amanda Stronza (2001) note that tourism has been relevant to anthropology because of its ability to affect almost all peoples of the world, its major economic importance, and its theme of cultural contact, but the subdiscipline has primarily channeled these interests into two main themes: the origins of tourism, with the focus on the tourist, and the impact of tourism, with the focus on local people. Stronza further notes that the subdiscipline comes with a certain set of key assumptions. One is that tourism is a "vector" from the tourist's point of origin to the point of visitation, and that this vector can bring a sense of imposition or oppression wherein local people become passive recipients who lose their culture as the price of participation or who commodify their culture to satisfy tourist expectations. Stronza critiques this assumption, hoping to inspire new questions and research objectives for the anthropology of tourism. Instead of the constant focus on local impacts, she asks us to consider why "hosts" participate in tourism and in what ways. She also wonders what actually motivates tourists to travel, and she desires greater attention to the analysis of the concept of tourism itself.

18. The tourism described by Dean MacCannell (1976)—the leisured Euro-Americans seeking escape from their superficial, modern existence—is also not quite what is at work here.

19. Island Solutions was subsequently rebranded under a new name in 2014.

20. The changing epistemic languages of the life sciences are central to current articulations of life and value under the sign of biocapital (what we might call ecobiocapital). Kaushik Sunder Rajan writes about biotech that "forms of corporate PR are now tied to the production of scientific fact which is supremely authoritative and is moreover in this case fact about 'life itself.' Therefore, simultaneous to exploring the rhetorical and discursive apparatus of corporations is the need to explore the sorts of scientific facts genomics provides," and, I would add, the facts contemporary Anthropocene logics, products, and practices provide. These facts point to a new "ensemble of techniques, practices, and institutional structures" for the Anthropocene (2006, 135).

21. This daily schedule was copied from a sign hanging on the wall of the Island Academy.

22. Scott (2004).

23. One version of "colonial model" comes from a conversation in the history of science about the development of the sciences in the colonial period. George Basalla (1967) has a three-phase typology of the spread of Western sciences throughout the world, one phase of which is the "colonial model." Mark Harrison (2005) notes that "Basalla posited a universal model for the diffusion of

Western science, from an initial phase of exploration—in which colonies provided raw data and materials for scientific analysis in the West—to formal colonial dependence and, ultimately, to independence." Venni Krishna (1992), discussing Basalla, describes in his historical investigation of British colonial India that one category within the colonial model is that of the colonial "gate keeping" scientists, the holders of authority who discriminated against Indian scientists and generally refused to grant Indian scientific work the designation of knowledge.

24. Braun (2002).

25. My loose definition of "classic" Caribbean scholarship spans the mid-twentieth century and marks the demarcation of the Caribbean as a visible scholarly region, a response to its prior invisibility and "non-place" status for social science because it was believed that there were no "original cultures" there (Trouillot 1992).

26. Much as Stronza (2001) notes that tourism is interesting for anthropologists in part because of its "laboratory situation" in terms of cultural contact, the Caribbean and its islands have long been a laboratory situation for anthropology and its ideas about social change. For example, in 1937 Melville J. Herskovits published *Life in a Haitian Valley,* his ethnological description of the routine "phases of life" in one rural Haitian village. The point here is that Haiti and the Caribbean became a testing ground for the development of the theory of acculturation; the region's history and its people became subjects of anthropological experimentation and investigation.

A decade later the Cuban sociologist and functionalist Fernando Ortiz (1947) published *Cuban Counterpoint,* a study of "the economic, social, and cultural aspects resulting from the interplay of influences between Africans and Latin Americans" (Malinowski 1922, ix). He hoped to critique the concept of acculturation in the region, developing an alternative notion of transculturation that moves beyond notions of conversion and submission to norms, promoting the coproduction of the "new reality of civilization" out of cultural contact. Cuba was configured as a socioecological amalgam in this investigation, and while Ortiz's terms and conclusions are dated when compared to the Anthropocene exemplifications made by the Island Academy and Island Systems in Eleuthera, there is a similar development of the field at work.

Finally, in his 1971 consideration of the development of Jamaican society, Edward K. Braithwaite developed the famous notion of creolization. He saw creolization as a "culture action" and social process through which strangers, to an environment and to one another, met in a situation of asymmetry and developed a peculiar "new" social construct. The Caribbean became the paradigmatic site to study creolizing phenomena, even as this concept was applied throughout the Americas and the world.

27. Scott (2004).

28. This statement is an allusion to Laarman and others (1989) work on tourism.

29. Raffles (2002, 158).

30. For Hugh Raffles ecological appeals to economic rationality depend on a shared view of evenly unfolding time and a desire for sustainability limited by notions of perpetual crisis, primitive accumulation, and particular cultural and

economic logics in partnerships with business. As an example, he states that people must desire forest-based futures to partner with ecological desires for forest conservation. Citing the Amazon Rainforest, he claims that this desire replaces the fetishization of wood as a commodity with the scientific fetishization of "tree-ness." Management projects, within this framing, "aim to produce a cosmopolitan tree with a localized meaning and specificity, a richly situated yet mobile identity" (2002, 160).

31. Raffles explains that such pragmatism comes from the "politics of negotiating multiple publics," and such negotiations imbue forms of nature with "transformative translocality." Forms of nature possess a translocality, which "creates anew those with whom it comes into contact," by interpolating them into a set of translocal conversations or debates (2002, 159). These conversations are driven by "crisis ridden rhetorics of biodiversity and habitat conservation, the combative confidence of the neoliberal assertion of entrepreneurial rights, and the authoritative expansion of natural scientific expertise into the realm of social policy" (160, following Hawkins 1993).

32. For Arun Agrawal there are three of what he calls positive modes of subject formation. These interdependent modes are subject formation through (1) practices of scientific inquiry that target specific types of subjects, (2) disciplinary practices of differentiation, and (3) practices of self-formation through thought. The understanding of subjectivity is the most important aspect of the analytic of environmentality, an analytic that, when combined with the shifting production of knowledge, politics, and institutions around environmental governance and management, becomes one specific "optic" for analyzing environmental politics (2005, 221).

3. SEA OF GREEN

1. A version of this chapter was published in *Environment and Society Advances in Research* in 2010.

2. Cleare (2007).

3. Pirates are here romantically portrayed as visitors as opposed to the more common portrayal of them as illegal smugglers (Adderley 2007).

4. Cleare (2007, 35).

5. Cleare (2007, 158).

6. The IUOTO became the United Nations World Tourism Organization (UNWTO) in 1975.

7. Cleare (2007, 211).

8. Cleare (2007, 233). Junkanoo is a Bahamian folk and musical tradition with roots in the slave period, but the Bahamian government has funded the event and influenced its promotion as a tourist attraction over the winter holiday season.

9. Islands of The Bahamas (n.d.).

10. Islands of The Bahamas (2016).

11. Cleare (2007, 19).

12. This group was created under Agenda 21 of the 1992 UN Conference on Environment and Development.

13. This hotel no longer exists as of approximately 2013, when it was shuttered to make way for the development of a new megaresort in the same location.

14. I was allowed to observe the workshop as a visiting U.S. graduate student, and I had learned of the meeting on a local informational website. I had to contact the requisite officer in the Ministry of Tourism to obtain access to the event.

15. UNWTO (2008).

16. See Post (2008).

17. Schneider et al. (2007).

18. Ghina (2003, 145).

19. Julca and Paddison (2009).

20. See "About" (n.d.).

21. Luky Adrianto and Yoshiaki Matsuda define small islands (a separate category from small-island states) as "islands with a population of 500,000 or less and a land area of approximately 10,000 km squared or less," although the SIDS Network does not follow this accounting (2002, 394).

22. Van der Velde et al. (2007).

23. See Moss et al. (2010); and Patz et al. (2005). In terms of the institutionalization of climate science as the defining issue of the Anthropocene, Amy Dalmedico and Helene Guillemot (2009) have compiled a brief and broad history of the genesis of the notion of climate change, rooting it in the post–World War II marriage of meteorology and new calculators for large-scale weather-forecasting systems. In the 1970s climate modeling converged with weather modeling, utilizing the development of new computers and satellites, and the scale of these scientific models became conceptualized and operationalized as global rather than regional. It became possible to imagine climate as an object with dynamic variables and physical processes based in the fluid mechanics of the atmosphere that could be calculated. Since its inception and prior to the "discovery" of climate change, climate modeling has always been about societal needs such as the assumed relevance of weather forecasts and the dynamic effects of climate on agriculture. But in the 1980s, following developments in GCS fields, such as geochemistry and paleoclimatology combined with climate modeling, "anthropogenic" climate change was born. The work of these earth scientists validated the models of the climate physicists, making them increasingly relevant as the exemplars for anthropogenic climate change, also known as "global warming."

In the 1980s the World Meteorological Organization, a specialized agency of the United Nations, designed its World Climate Programme, which hosted meetings that would lead to the creation of the international expert organization, the IPCC, whose scientists review policies related to greenhouse gas emissions and adaptation and mitigation strategies for the projected effects of global warming. Dalmedico and Guillemot write that "climate change has been transformed from a complex scientific topic to a political issue with national, economic, social, and diplomatic ramifications involving conflicting economic interests, conceptions of law and equity, and perceptions of the future." They also note that the IPCC is therefore particularly interesting as an international body because its evaluation process is controlled by scientists (2009, 212).

Despite fears that their results are not adequately understood or appreciated, GCS scientists are construed as essential experts having the appropriate knowledge to inform sound policy. It is therefore crucial to understand that science-based development tools involved in adaptation and mitigation measures and the creation of vulnerability indices are a major driver of the stakes when it comes to internationally recognized small-island issues today.

24. Hamilton (2003).

25. Kelman (2010a).

26. In this literature small-island states are construed as being in worse shape than they were before they became a part of the global economy in that the "traditional practices" of island people are seen as more sustainable than contemporary livelihood practices, especially in terms of agriculture and water use or "living in harmony" with the environment; see, for example, Ghina (2003); and van der Velde et al. (2007). For a general discussion of development, disaster, and vulnerability, see Cuny (1983).

27. See chapter 1, in this volume.

28. McConney (2002). For example, one area of international concern is small-island state fisheries management, and, as a result of this focus, social science is increasingly oriented around developing an understanding of fisher behavior and livelihood strategies. These are incorporated into the creation of more holistic models that include human dimensions and link individual decisions to state regulations and international recommendations (see chapter 4, in this volume and Colburn et al. 2006).

29. There are currently a number of anthropological studies concerning local adaptation to climate change around the world; see, for example, Adger and Kelly (1999); and Berkes and Jolly (2001). Nelson, West, and Finan (2009) describe climate change adaptation as multiscalar, linked to socioeconomic inequalities, interrelated with natural and global stresses, and requiring community participation. Anthropology, they argue, is disciplinarily suited to engaging with these phenomena because it can contribute to an understanding of human dimensions and focus attention on local vulnerabilities, that is, the "social aspects" of adaptation that may or may not fit into current policy frameworks. These authors call for increased anthropological inclusion and for participation in the development of the assessment reports of the IPCC.

30. See Lahsen (2005); and Yearley (2009). Nicole Peterson and Kenneth Broad attend to what they call "climate narratives" created by climate scientists and the way that these narratives generate notions of globality and allow for the creation of new scientific productions. For these two authors anthropology plays a checks-and-balances role, defending local priorities in climate change issues: "Anthropologists may likely find them-selves arguing against the importance of global warming as a major risk factor versus more immediate (and longstanding) drivers of vulnerability including property rights, education, and access to water and health care. Climate change discourse has the potential to obfuscate unequal power relations, letting governments off the hook for poor environmental and social policies and practices" (2009, 80–81). Roncoli, Crane, and Orlove (2009) argue that anthropological contributions to the field can go beyond structured surveys and analyses based on quantified results.

31. Rabinow (2008).
32. Cuny (1983); Pelling and Uitto (2001).
33. Lazrus (2009).
34. McCarthy et al. (2001, 6).
35. Schneider et al. (2007, 782). Lino Briguglio (2000) has also defined the term *vulnerability* as the potential for aspects of a system to be damaged by external impacts.
36. The UN Framework Convention on Climate Change also promotes information development, the creation of resource inventories, capacity-building priorities, systemic models, and the development of decision-making tools for SIDS.
37. One of the earliest discussions of economic vulnerability indices for small islands came from Briguglio (1995), following the call for increased vulnerability calculations from the Barbados Program of Action in 1994, and one of the latest institutionalizations of the current expanded notion of vulnerability indices considering climate change comes from Mimura et al. (2007).
38. Ghina (2003).
39. Climate change is even connected to the vulnerability of small islands through the migration of the wealthier and skilled sectors of the population who can afford to leave when island life becomes difficult. This can lead to remittance dependency and a lack of qualified people remaining in the island country of origin (Julca and Paddison 2009).
40. Adrianto and Matsuda (2002); Costanza, Segura Bonilla, and Martínez Alier (1996).
41. Ghina (2003); Hamilton (2003). With regard to the Maldives, Fathimath Ghina has written that "the inherent nature of the islands predispose[s] them to frequent damage from storm and wave surges. This vulnerability has direct implications for a number of activities related to the economy and indeed livelihoods" (2003, 144).
42. London (2004).
43. R. Moore (2002).
44. Julca and Paddison (2009).
45. These categories are intended to represent social inequities, the lack of economic diversification, and various environmental vulnerabilities to climate impacts.
46. The goal is to generate "political will" toward adaptive action and to create economic incentives for adaptation measures. There are those who argue that small islands must be made into global examples for effective climate change planning in the Anthropocene because they are characterized as some of the most vulnerable places in the world. Some projects attempt to quantify and value the impacts of climate change to make them economically comprehensible for policy makers (Ghina 2003). Other experts recommend funding so called no-regret projects in vulnerable islands, whereby the environment and economic sectors both "win," as well as developing a valuation scheme to represent the general national costs of climate change (London 2004).
47. Barnett, Lambert, and Fry (2008).
48. Lazrus (2009).

49. For Heather Lazrus vulnerability is a "matter of representation to which questions of agency are central" (2009, 240).

50. J. Campbell (1997); Lazrus (2012).

51. Barnett and Adger (2003).

52. Barnett, Lambert, and Fry (2008) also offer a thoughtful history and critique of the Environmental Vulnerability Index.

53. Epeli Hau'ofa (2008) leads the way by invoking the "Sea of Islands" argument about Pacific island connectivity, mobility, knowledge, and resilience in the face of both climate change and reductive summations of island isolation and vulnerability (see also Lazrus 2012). This is a local framework that can begin to resist the more limiting aspects of the socioecologic of vulnerability.

54. A. Moore (2010).

55. Strachan (2002).

56. Anna Tsing (2005) has critiqued climate science, stating that climate change creates a form of "global nature" through experts' strict adherence to climate models on a global scale in which the local disappears, but in this case the local is instead transformed and homogenized.

57. Coombe (1998).

58. A. Moore (2015b).

59. Bethel (2013, para. 2).

60. See Lazrus (2012); and Malm (2006).

61. Oliver-Smith (2009).

62. Luhmann (1998).

4. AQUATIC INVADERS IN THE ANTHROPOCENE

1. A version of this chapter was published in *Cultural Anthropology* (A. Moore 2012). This chapter owes a great deal to the work of Bonnie McCay and Gisli Palsson and to the maritime anthropological journal *MAST*. See also van Ginkel and Verrips (2002).

2. Green and Cote (2009); Morris and Akins (2008).

3. Albins and Hixon (2008).

4. See Buchan (2000). Nassau grouper are one of the "holy trinity" of Bahamian commercial species, along with queen conch and spiny lobster (crawfish).

5. Chaplin (2006); Chiappone et al. (2000).

6. Maurer (2000) describes a fish story as a critique of ontology and teleology guided by a fish species.

7. Maurer (1997).

8. See Helmreich (2003); Palsson (1991); Palsson and Durrenberger (1990); M.E. Smith (1977); and Spoehr (1980).

9. See McCay (2001). In this discipline notions of cultural forms of sea tenure, fishing knowledge, and the anthropological fight to refute the vision of the ocean as a universal commons, free for corporate or community exploitation, have been especially important (McCay and Acheson 1987). Maritime studies have also been seen as strong counterpoints to the "tragedy of the commons" argument (Hardin 1968; modified by Feeny et al. 1990).

10. McCay (2008).

11. McCay (1978, 397).

12. See Flaaten and Heen (2007). McCay (2001) notes that modern fisheries management, a development in the Global North since the 1960s, is intended to be a rational, science-based system of control by the state with special power over aquatic common pool resources. I note that social and cultural factors are seen as increasingly important within forms of cutting-edge scientific management.

13. Palsson (1991, 23).

14. Marcus (2002).

15. Stoffle and Minnis (2007).

16. See chapter 1, in this volume, for more on the Marine Reserve Network process.

17. Hawkins and Roberts (2004); Himes (2007).

18. It is crucial to keep exploring how fisher practices are targeted, interpreted, and managed in Anthropocene institutional relationships and rhetoric as well as through relations of socioecological encounter. Maritime anthropology has already done a great deal of work to dispel antiquated assumptions about fishing life and to engender an awareness of the unstable positions of fishers in contemporary society.

19. I follow Helmreich (2009), who has taken his own maritime interest into the realm of marine microbial science and genetic prospecting.

20. For example, anthropologists are frequently involved in the project of *explaining* island villagers to governmental fisheries developers who want them to become fishers in expanding fisheries so that they can participate in capitalist markets and diversify island economies away from low-income crops or subsistence (e.g., see Rodman 1987 on Melanesia).

21. Colburn et al. (2006, 232). Anthropologists at the National Marine Fisheries Service examined the impact of regulation on community organization and function, and anthropology became a full-blown program there in the 1990s.

22. See Clay and McGoodwin (1995).

23. On community, see Agrawal (2005); Brosius, Tsing, and Zerner (1998); and West (2006).

24. For an example of differences within a Caribbean fishing community, see Doyon (2007).

25. LaFlamme (1979).

26. Dolphin is also known as mahi mahi.

27. Gascoigne (2002); Chiappone (2000); Clark et al. (2005).

28. Jackson et al. (2001, 629). Overfishing has even retroactively been "discovered" in the archaeological record (Blick 2007).

29. U. Bell (2004); Holling, Gunderson, and Ludwig (2002); Pauly (2006).

30. Bahamian fisheries are often touted as much more healthy and economically viable than the majority of Caribbean fisheries, and therefore a lot of the Anthropocene discourse around marine protection in The Bahamas is precautionary rather than a response to "actual" crisis conditions.

31. In contrast, invasive-species flyers designed by the Bahamas Environment Science and Technology Commission around the turn of the millennium contained no images of marine species, just close-ups of green invasive plants, a cute juvenile raccoon, and a passively perched Eurasian dove.

32. Sundaravadanan (2009).

33. Head (2007).

34. Stefan Helmreich (2009, 150) also notes that there is a specific Hawaiian politics of invasion rooted in indigenous politics and what he calls a particular "archipelagic imagination" of the sea as mobility—indigeneity in Hawaii means being both native and mobile.

35. Helmreich (2005); Subramaniam (2001).

36. As Jean and John Comaroff explain through a discussion of plant biology and management in South Africa, disaster or crisis around invasive species highlights "conditions of being" in postcolonial nation-states (2001, 629).

37. Lightbourn (2000).

38. Bethel (2008).

39. See also Kirksey and Helmreich (2010).

40. Celia Lowe investigates Indonesian biologists' scientific designations of species and how the designation of species' endemism, in this case of a particular primate, is used to anchor regional biodiversity conservation agendas. The designation of species with specific biological characteristics is a political act with implications for promoting the legitimacy of Indonesian conservation biology and expertise; the strategic essentialism of certain creatures as endemic satisfies multiple purposes for both Indonesian scientists and the state. Endemic organisms thus become "metaphorical 'keystone species' for Indonesian conservation biology" (2006, 52).

41. Cori Hayden's work on brine shrimp in plant molecular bioprospecting in Mexico is another example of the way animals have been shown to be essential actors in scientific networks and narratives. The tiny brine shrimp used in chemists' tests for plant potency have become agents of value in that their bodies reveal if a plant is potentially valuable or not through the act of living or dying in solution. These shrimp are gate-keeping actors and brokers of value because of their specific characteristics as monetarily and ethically cheap model organisms for the biosciences. By using the shrimp, Mexican plant chemists can make a legitimate claim to patentable property they did not historically have in international drug discovery circles. The shrimp embody "the representational capacities of scientific research conducted in the name of promises to wider access to biodiversity-derived value" (2003, 192).

42. See Haraway (2008, 5).

43. West (2006). For a description of sea turtles as "flagship species," see Eckert and Hemphill (2005); and Frazier (2005).

44. Fuentes (2010).

45. D. Campbell (1978b, 56).

46. Cox (1999).

47. Albins and Hixon (2008, 237).

48. Whitty (2009, 64).

49. This move is also becoming popular in U.S. management circles, and, with the support of marine conservation nongovernmental organizations, the fish is promoted in the Florida Keys and New York as a boutique meal (Ferguson and Akins 2010; Rosenthal 2011).

50. "Over 800 Invasive Lionfish Are Caught in Half a Day," *Tribune,* July 20, 2009, A10.

51. C. Thompson (2010).

52. Galanis (2016).

53. Aronson and Precht (2006); Lowe (2006); Mora and Sale (2011); Paddack et al. (2009); Schiermeier (2004).

54. For social and ecological complexity, see Anderies, Janssen, and Ostrom (2004). For Panarchy, see Holling, Gunderson, and Ludwig (2002). For simplification, see Bromley (2008); West (2006); and chapters 1 and 2, in this volume.

55. Palsson (1994, 921).

56. Segal (2001).

5. DOWN THE BLUE HOLE

1. This chapter is written in honor of the professional divers Wes Skiles and Agnes Milowka, who participated in the projects described here and who have since passed away in separate diving accidents.

2. NOVA (2010).

3. The archaeologist was touted as the first Bahamian to dive into one of the most unexplored blue holes. The anthropologist is a mentor of mine, and it was he who invited me to participate, knowing that I was interested in the subject.

4. G. Moore (1964, 3).

5. All this information has been taken directly from what was reported on the television show NOVA (2010).

6. Raffles (2002).

7. UNESCO World Heritage Center (2014).

8. Schwabe at al. (2006).

9. Schwabe et al. (2007); Poucher and Copeland (2006).

10. Keen (2008).

11. Bahamas Information Services (2006).

12. This has not been officially demonstrated, but is often speculated in blue hole circles.

13. All the larger islands of the northern Bahamas have stands of Caribbean pine forest, though not all these forests are as large as those in Andros and Abaco.

14. We were also told about a half-shark, half-octopus creature called the lusca, said to make the holes bubble and "breathe" and to devour swimmers and divers. But the only people who mentioned this were foreigners who worked in dive shops catering to tourists and research visitors, though they described the lusca as though it was a widespread local legend.

15. Carnegie (2002); Khan (2004); Thomas (2004).

16. Forte (2005).

17. See Forte (2005, 25). This is not quite what Maximilian Forte describes in Trinidad, where the promotion of the indigenous Carib identity, with its concomitant narration of a primordial history from a pre-Columbian past to the postcolonial present, is a project of reappropriating a silenced history,

reviving an "extinct" representation of identity, and promoting a political nationalism. The promotion of an indigenous Carib identity in Trinidad is unusual, considering most anthropological and historical literature on the region, including work in Trinidad, has espoused the fact that the original Amerindian inhabitants of the region were all removed or killed soon after Europeans' arrival. This notion of erasure is one of several that supposedly uphold the uniqueness of the Caribbean region in the world, a region with no aboriginal populations, a region whose history has been imagined as one of migration and foreign encounter with equally foreign others. The popular promotion of a creolized Caribbean implicitly denies the existence of any one pure and linear Caribbean heritage, and the figure of the Amerindian has become a symbolic force in Trinidad that challenges or supersedes this view as an alternative claim to creole territorial legitimacy.

18. V. Young (1993).

19. L. Braithwaite (1957); E. Braithwaite (1971); M.G. Smith (1965).

20. This is in stark contrast to the "multiple metaphorics of mixing," of which creole is a part, that make up the Caribbean region for authors such as Aisha Khan (2001, 2004).

21. This thought dovetails with Bill Maurer's work on ideas about nature, kinship, and citizenship in the British Virgin Islands. If, in Caribbean places like the Virgin Islands and The Bahamas, national identity is largely homogenous and supposedly based on a shared sense of history in which there are no "indigenous" rights or ties to the nation, then what other narratives ground and literally naturalize the presence of people within a national boundary? Maurer writes about the way a creolized "network of technologies" has come to take the place of this ground in the Virgin Islands, technologies that come to stabilize "'ethnic' stereotypes; 'races' and communities bound to 'places'; 'families' and 'genealogies'; 'land' and 'country'; 'classes' and 'parties'; 'states and 'societies'; 'individuals' who owe nothing to society, and 'nations'" (1997, 263–64). His attention is to the legal systems that reify certain understandings of blood and kinship as a tie to land and citizenship rights.

22. Stuart Hall argues that Caribbean identity has two modalities: being/oneness and becoming/ discontinuity. The first notion of identity for Hall is a sort of fiction that "is the truth, the essence, of 'Caribbeanness,'" which underlies the more superficial differences within the region and which gains its social usefulness from its "rediscovery" by various Caribbean leaders (1994, 393). The second notion of identity is the opposite of the first in that it stands as the other side of oneness—it is the recognition of "the ruptures and discontinuities which constitute, precisely, the Caribbean's 'uniqueness'" (394).

23. Often it is assumed that self-described multicultural or multiethnic small nations struggle for a cohesive national identity, but the reasons for why an imagined national unity is necessary are unclear, beyond the fact that some think this lends itself to political stability. But when we examine national narratives, questions about whether Bahamians intrinsically *need* a historically enriched national sense of identity become irrelevant.

24. Knudson et al. (2008).

25. See Sommer and Carrier (2010).

26. Kohler (2002, 32).

27. See, for example, the phenomenon of geotourism (Newsome and Dowling 2010).

28. As of the spring of 2016, the AMMC had completed the nomination form for these blue hole regions, although they could potentially declare all blue holes as "heritage." There is a two-year nomination process under the UNESCO World Heritage Convention, but the prime minister will eventually be the final arbiter for what is submitted as a possible site within the country.

29. His name has been changed.

30. See Rose and Novas (2005) and Olson (2010).

31. For a discussion of social nature in the lab, see Knorr Cetina (1999).

32. Wise and Moore (2013).

33. See also West and Carrier (2004).

34. Discussion with the author, Abaco, Bahamas, August, 2009. This microbiologist examines extreme environments on the modern earth and extrapolates these as reflecting the biota and geochemistry of the early earth because "microbes don't leave very good fossils."

CONCLUSION

1. See Latour (2014).

2. Meyer (2018).

3. Jason Moore has proposed "Capitalocene" to more accurately reflect the fact that capitalism itself is an ecology—a "way of organizing nature" (2016, 7). Donna Haraway (2016) has proposed "Chthulucene" to more evocatively express another kind of story in which Anthropos, capital, technocracy, and the scientific modern synthesis are decentered in favor of more entangled, messy, and earthly narratives about pasts, presents, and, most important, potential futures. These are perhaps the two most well-known alternative terms for the Anthropocene, but there are many other possible labels (see also Howe and Pandian 2016).

4. A. Moore (2016).

5. Marcus (1995).

6. Rabinow (1989); Kelty et al. (2006).

7. Anthropocene logics are components of positional self-knowledge and self-fashioning; that is, they lend themselves to evolving forms of sociality and citizenship. Emergent alliances are created to shape these subjectivities every day, transnationally and locally, from the large-scale climate meetings in places like Paris or Copenhagen to weekly gatherings of students in New Providence for field trips with local nongovernmental organizations. Yet subjects are not easily reproduced. It remains to be seen not if but *how* the contingent processes of the Anthropocene will manifest in forms of self-fashioning in The Bahamas, the small islands of the Caribbean, and around the world. These will not be ubiquitous environmental subjects (Agrawal 2005).

8. Bennett (2010); Braun and Whatmore (2010).

9. Meyer (2018).
10. A. Moore (2015b).
11. Nixon (2015); Bethel (2017); Galanis (2017).
12. Stengers (2010).
13. Gibson-Graham (2011, 19).

References

"About." n.d. Accessed February 18, 2019. http://aosis.org/about/.http://aosis.org/about/.

Adderley, Paul L. 2007. Foreword to Cleare, *History of Tourism*, 17–18.

Adger, W. Neil, and P. Mick Kelly. 1999. "Social Vulnerability to Climate Change and the Architecture of Entitlements." *Mitigation and Adaptation Strategies for Global Change* 4 (3–4): 253–66.

Adrianto, Luky, and Yoshiaki Matsuda. 2002. "Developing Economic Vulnerability Indices of Environmental Disasters in Small Island Regions." *Environmental Impact Assessment Review* 22 (4): 393–414.

Agard-Jones, Vanessa. 2013. "Bodies in the System." *Small Axe* 17 (3): 182–92.

Agrawal, Arun. 2005. *Environmentality: Technologies of Government and the Making of Subjects*. Durham, NC: Duke University Press.

Albins, Mark A., and Mark A. Hixon. 2008. "Invasive Indo-Pacific Lionfish *Pterois volatins* Reduce Recruitment of Atlantic Coral-Reef Fishes." *Marine Ecology Press Series* 367:233–38.

Anderies, John M., Marco A. Janssen, and Elinor Ostrom. 2004. "A Framework to Analyze the Robustness of Social-ecological Systems from an Institutional Perspective." *Ecology and Society* 9 (1): 1–17.

Aronson, Richard B., and William F. Precht. 2006. "Conservation, Precaution, and Caribbean Reefs." *Coral Reefs* 25:441–50.

Bacon, Francis. 1624. *The New Atlantis: A Work Unfinished*. United Kingdom.

"Bahamas." 2015. *Atlas of Marine Protection*. September 4, 2015. www.mpatlas.org/region/country/BHS/.

Bahamas Information Services. 2006. "Antiquities Board Gets New Mandate." Accessed February 26, 2019. www.bahamas.gov.bs/bahamasweb2/home

.nsf/vContentW/9B43DE2FC68D86D8852571DC00728D60!OpenDocument&Highlight = 0,AMMC.

Bahn, Paul, and John R. Flenley. 1992. *Easter Island, Earth Island*. London: Thames and Hudson.

Barnes, Jessica, Michael Dove, Myanna Lahsen, Andrew Mathews, Pamela McElwee, Roderick McIntosh, Frances Moore, et al. 2013. "Contribution of Anthropology to the Study of Climate Change." *Nature Climate Change* 3:541–44.

Barnett, Jon. 2002. "Rethinking Development in Response to Climate Change in Oceania." *Pacific Ecology* 1. www.pacificecologist.org/archive/pe1.html.

Barnett, Jon, and W. Neil Adger. 2003. "Climate Dangers and Atoll Countries." *Climatic Change* 61 (3): 321–37.

Barnett, Jon, Simon Lambert, and Ian Fry. 2008. "The Hazards of Indicators: Insights from the Environmental Vulnerability Index." *Annals of the Association of American Geographers* 98 (1): 102–19.

Barry, Andrew, and Georgina Born. 2013. *Interdisciplinarity: Reconfigurations of the Social and Natural Sciences*. London: Routledge.

Basalla, George. 1967. "The Spread of Western Science." *Science* 156 (3775): 611–22.

Bell, Hugh M. 1943. *Bahamas: Isles of June*. London: Williams and Norgate.

Bell, Udy. 2004. "Overfishing: A Threat to Marine Biology." *UN Chronicle* 41 (2): 17.

BenDor, Todd, Douglass A. Shoemaker, Jean-Claude Thill, Monica A. Dorning, and Ross K. Meentemeyer. 2014. "A Mixed-Methods Analysis of Social-Ecological Feedbacks between Urbanization and Forest Persistence." *Ecology and Society* 19 (3): 3.

Benitez-Rojo, Antonio. 1996. *The Repeating Island: The Caribbean and the Postmodern Perspective*. Durham, NC: Duke University Press.

Bennett, Jane. 2010. "Thing Power." In Braun and Whatmore 2010, 35–62.

Berkes, Fikret. 2003. "Rethinking Community-Based Conservation." *Conservation Biology* 18 (3): 621–30.

Berkes, Fikret, and Dyanna Jolly. 2001. "Adapting to Climate Change: Social-Ecological Resilience in a Canadian Western Arctic Community." *Conservation Ecology* 5 (2): 18–33.

Bethel, Nicolette. 2008. *Bahamian Essays Originally Published in the Nassau Guardian*. Vol. 1 of *Essays on Life*. Morrisville, NC: Lulu.

———. 2013. "The New Colonialism." *Blogworld on Wordpress* (blog). https://nicobethelblogworld.wordpress.com/2013/12/12/the-new-colonialism-or-things-fall-apart-sp-downgrade-if-no-tax-follow-through-the-tribune/.

———. 2017. *Blogworld on Wordpress* (blog). Accessed February 26, 2019. https://nicobethelblogworld.wordpress.com/.

Blick, Jeffrey P. 2007. "Pre-Columbia Impact on Terrestrial, Intertidal, and Marine Resources, San Salvador, Bahamas (A.D. 950–1500)." *Journal for Nature Conservation* 15:174–83.

Blue and Green Tomorrow Communications. 2013. *The Guide to Sustainable Tourism*. Accessed February 26, 2019. www.blueandgreentomorrow.com.

Boruff, Bryan J., and Susan L. Cutter. 2007. "The Environmental Vulnerability of Caribbean Island Nations." *Geographical Review* 97 (1): 24–45.
Braithwaite, Edward K. 1971. *The Development of Creole Society in Jamaica, 1770–1820*. Oxford: Clarendon.
Braithwaite, Lloyd. 1957. "Sociology and Demographic Research in the Caribbean." *Social and Economic Studies* 6 (4): 523–71.
Braun, Bruce. 2002. *The Intemperate Rainforest: Nature, Culture, and Power on Canada's West Coast*. Minneapolis: University of Minnesota Press.
Braun, Bruce, and Sarah Whatmore, eds. 2010. *Political Matter: Technoscience, Democracy, and Public Life*. Minneapolis: University of Minnesota Press.
Briguglio, Lino. 1995. "Small Island Developing States and Their Economic Vulnerabilities." *World Development* 23 (9): 1615–32.
———. 2000. "An Economic Vulnerability Index and Small Island Developing States: Recent Literatures." Working Paper Seminar on Islands Studies, Kagoshima University, Pacific Island Studies Center.
Bromley, Daniel W. 2008. "The Crisis in Ocean Governance: Conceptual Confusion, Spurious Economics, Political Indifference." *MAST* 6 (2): 7–22.
Brosius, J. Peter, Anna L. Tsing, and Charles Zerner. 1998. "Representing Communities: Histories and Politics of Community-Based Natural Resource Management." *Society and Natural Resources* 11:157–68.
Buchan, Kenneth C. 2000. "The Bahamas." *Marine Pollution Bulletin* 41 (1–6): 94–111.
Campbell, David G. 1978a. *The Ephemeral Islands: A Natural History of The Bahamas*. London: Caribbean Press.
———. 1978b. "The Invasion of the Exotics." In Campbell 1978a, 56–66.
Campbell, John. 1997. "Examining Pacific Island Vulnerability to Natural Hazards." In *Proceedings, VIII Pacific Science Intercongress*, edited by A. Planitz and J. Chung, 53–62. Suva, Fiji: South Pacific Programme Office, United Nations Department for Humanitarian Affairs.
Campbell, Timothy C., and Adam Sitze. 2013. *Biopolitics: A Reader*. Durham, NC: Duke University Press.
Carnegie, Charles V. 2002. *Postnationalism Prefigured: Caribbean Borderlands*. New Brunswick: Rutgers University Press.
Caro, Tim, Jack Darwin, Tavis Forrester, Cynthia Ledoux-Bloom, and Caitlin Wells. 2011. "Conservation in the Anthropocene." *Conservation Biology* 26 (1): 185–88.
Carrier, James, and Donald Macleod. 2005. "Bursting the Bubble: The Sociocultural Context of Ecotourism." *Journal of the Royal Anthropological Institute* 11 (2): 315–34.
Cepek, Michael L. 2011. "Foucault in the Forest: Questioning Environmentality in Amazonia." *American Ethnologist* 38 (3): 501–15.
Chakrabarty, Dipesh. 2009. "The Climate of History: Four Theses." *Critical Inquiry* 35:197–222.
———. 2012. "Postcolonial Studies and the Challenge of Climate Change." *New Literary History* 43:1–18.
———. 2013. "History on an Expanded Canvas: The Anthropocene's Invitation." *Anthropocene Project Keynote Presentation: HKW*. Accessed

February 26, 2019. www.hkw.de/en/programm/projekte/veranstaltung/p_83957.php.

Chaplin, Gordon. 2006. "A Return to the Reefs." *Smithsonian* 36 (11): 41–48.

Chiappone, Mark, Robert Sluka, and Kathleen Sealey. 2000. "Groupers (Pisces: Serranidae) in Fished and Protected Areas of the Florida Keys, Bahamas, and Northern Caribbean." *Marine Ecology Press Series* 198:261–72.

Clark, S.A., A.J. Danylchuk, and B.T. Freeman. 2005. "The Harvest of Juvenile Queen Conch *(Strombus gigas)* Off Cape Eleuthera, Bahamas: Implications for the Effectiveness of a Marine Reserve." *Proceedings of the Gulf and Caribbean Fisheries Institute* 56:705–13.

Clay, Patricia M., and James R. McGoodwin. 1995. "Utilizing Social Sciences in Fisheries Management." *Aquatic Living Resources* 8 (3): 203–7.

Cleare, Angela B. 2007. *History of Tourism in The Bahamas: A Global Perspective*. Philadelphia: Xlibris.

Clifford, James. 1988. *The Predicament of Culture: Twentieth-Century Ethnography, Literature, and Art*. Cambridge, MA: Harvard University Press.

Clifford, James, and George Marcus, eds. 1986. *Writing Culture: The Poetics and Politics of Ethnography*. Berkeley: University of California Press.

Colburn, Lisa L., Susan Abbott-Jamieson, and Patricia M. Clay. 2006. "Anthropological Applications in the Management of Federally Managed Fisheries: Context, Institutional History, and Prospectus." *Human Organization*. 65 (3): 231–39.

Colwell, Rita. 1998. "Balancing Biocomplexity of the Planet's Living Systems: A Twenty-First Century Task for Science." *Bioscience* 48 (10): 786–87.

Comaroff, Jean, and John L. Comaroff. 2001. "Naturing the Nation: Aliens, Apocalypse, and the Postcolonial State." *Journal of Southern African Studies* 27 (3): 627–51.

Committee on the Evaluation, Design, and Monitoring of Marine Reserves and Protected Areas in the United States. 2001. *Marine Protected Areas: Tools for Sustaining Ocean Ecosystems*. Washington, DC: National Academy Press.

Conca, James. 2014. "The Anthropocene Part 1: Tracking Human-Induced Catastrophe on a Planetary Scale." *Forbes*, August 16, 2014. www.forbes.com/sites/jamesconca/2014/08/16/the-anthropocene-part-1-tracking-human-induced-catastrophe-on-a-planetary-scale/#ddoed17c2437.

Coombe, Rosemary. 1998. *The Cultural Life of Intellectual Properties: Authorship, Appropriation, and Law*. Durham, NC: Duke University Press.

Costanza, Robert, Olman Segura Bonilla, and Juan Martínez Alier, eds. 1996. *Getting Down to Earth: Practical Applications of Ecological Economics*. Washington, DC: Island.

Cousteau, Jean-Michel, and Jaclyn Mandoske. 2014. "Canaries of Climate Change: The Plight of Pacific Island Nations." *Jean-Michel Cousteau's Ocean Futures Society*, August 8, 2014. www.oceanfutures.org/news/blog/canaries-climate-change-plight-pacific-island-nations.

Cox, George W. 1999. *Alien Species in North America and Hawaii: Impacts on Natural Ecosystems*. Washington, DC: Island.

Crate, Susan A. 2011. "Climate and Culture: Anthropology in the Era of Contemporary Climate Change." *Annual Review of Anthropology* 40:175–94.

Crate, Susan A., and Mark Nuttall, eds. 2009. *Anthropology and Climate Change: From Encounters to Actions.* Walnut Creek, CA: Left Coast.
Crutzen, Paul J. 2002. "Geology of Mankind." *Nature,* 415.
Crutzen, Paul J., and Will Steffen. 2003. "How Long Have We Been in the Anthropocene Era?" *Climatic Change* 61 (3): 251–57.
Crutzen, Paul J., and Eugene F. Stoermer. 2000. "The 'Anthropocene.'" *Global Change Newsletter* 41:17–18.
Cuny, Frederick C. 1983. *Disasters and Development.* Edited by Susan Abrams. Oxford: Oxford University Press.
Dalmedico, Amy D., and Helene Guillemot. 2009. "Climate Change: Scientific Dynamics, Expertise, and Geopolitical Challenges." In Mallard, Paradeise, and Peerbaye 2009, 195–221.
Dalsgaard, Steffen. 2013. "The Commensurability of Carbon: Making Value and Money Out of Climate Change." *Hau* 3 (1): 80–98.
Darwin, Charles. 1839. *The Voyage of the Beagle.* London: Penguin Books.
———. 1859. *On the Origin of Species.* London: Penguin Books.
Davidov. Veronica. 2012. "From a Blind Spot to a Nexus: Building on Existing Trends in Knowledge Production to Study the Copresence of Ecotourism and Extraction." *Environment and Society: Advances in Research* 3:78–102.
Deloughrey, Elizabeth. 2013. "The Myth of Isolates: Ecosystem Ecologies in the Nuclear Pacific." *Cultural Geographies* 20 (2): 167–84.
Doyon, Sabrina. 2007. "Fishing for the Revolution: Transformations and Adaptations in Cuban Fisheries." *MAST* 6 (1): 83–108.
Eckert, Karen L., and Arlo H. Hemphill. 2005. "Sea Turtles as Flagships for Protection of the Wider Caribbean Region." *MAST* 3 (2) and 4 (1): 119–43.
Farbotko, Carol. 2010. "Wishful Sinking: Disappearing Islands, Climate Refugees and Cosmopolitan Experimentation." *Asia Pacific Viewpoint* 51 (1): 47–60.
Farnsworth, Stephen J., and S. Robert Lichter. 2011. "The Structure of Scientific Opinion on Climate Change." *International Journal of Public Opinion Research* 24 (1): 93–103.
Feeny, David, Fikret Berkes, Bonnie J. McCay, and James M. Acheson. 1990. "The Tragedy of the Commons: Twenty-Two Years Later." *Human Ecology* 18 (1): 1–19.
Ferguson, Tricia, and Lad Akins. 2010. *The Lionfish Cookbook: The Caribbean's New Delicacy.* Key Largo, FL: Reef Environmental Education Foundation.
Fiske, Shirley J. 2009. "Global Change Policymaking from Inside the Beltway: Engaging Anthropology." In Crate and Nuttall 2009, 277–91.
Flaaten, Ola, and Knut Heen. 2007. "Spatial Employment Impacts of Fisheries Management." *Fisheries Research* 85 (1–2): 74–83.
Forte, Maximilian. 2005. *Ruins of Absence, Presence of Caribs: (Post)Colonial Representations of Aboriginality in Trinidad.* Gainesville: University of Florida Press.
Franklin, Sarah, and Margaret Lock. 2003. *Remaking Life and Death: Toward an Anthropology of the Biosciences.* Santa Fe, NM: School for American Research Press.

Frazier, John. 2005. "Marine Turtles: The Role of Flagship Species in Interactions between People and the Sea." *MAST* 4 (1): 5–38.

Fuentes, Agustın. 2010. "Naturalcultural Encounters in Bali: Monkeys, Temples, Tourists and Ethnoprimatology." *Cultural Anthropology* 25 (4): 600–624.

Fujimura, Joan H. 2011. "Technobiological Imaginaries: How Do Systems Biologists Know Nature?" In *Knowing Nature: Conversations at the Intersection of Political Ecology and Science Studies,* edited by Mara J. Goldman, Paul Nadasdy, and Matthew Turner. Chicago: University of Chicago Press.

Furnass, Bryan. 2012. "From Anthropocene to Sustainocene: Challenges and Opportunities." Public Lecture, Australian National University, March 21, 2012.

Galanis, Tamika, dir. 2016. *When the Lionfish Came.* The Bahamas: privately produced.

———. 2017. "When the Lionfish Came." Accessed February 26, 2019. www.tamikagalanis.com/.

Galaz, Victor. 2012. "Geo-engineering, Governance, and Social-Ecological Systems: Critical Issues and Joint Research Needs." *Ecology and Society* 17 (1): 24.

Galvin, Kathleen A. 2009. "Transitions: Pastoralists Living with Change." *Annual Review of Anthropology* 38:185–98.

Gascoigne, Jo. 2002. "Grouper and Conch in The Bahamas: Extinction or Management? The Choice Is Now." Accessed 2002. www.macalister-elliott.com/media/reports/1651R03A.pdf (site discontinued).

Ghina, Fathimath. 2003. "Sustainable Development in Small Island Developing States: The Case of the Maldives." *Environment, Development and Sustainability* 5 (1–2): 139–65.

Gibson-Graham, J.K. 2011. "A Feminist Project of Belonging for the Anthropocene." *Gender, Place, and Culture: A Journal of Feminist Geography* 18 (1): 1–21.

Government of The Bahamas. 2015. *Intended Nationally Determined Contribution (INDC) under the United Nations Framework Convention on Climate Change.* November 17, 2015. www4.unfccc.int/sites/submissions/INDC/Published%20Documents/Bahamas/1/Bahamas%20INDC%20Submission.pdf.

Gowdy, John, and Lisi Krall. 2013. "The Ultrasocial Origin of the Anthropocene." *Ecological Economics* 95:137–47.

Graziadei, Daniel, and Johannes Riquet. 2015. "Writing Islands in the Anthropocene: Literature, Cultural Geography, and the De(con)struction of Islands." Paper presented at the Royal Geographical Society Annual Conference, Exeter, UK, September 2015. http://conference.rgs.org/AC2015/247.

Green, Stephanie J., and Isabelle M. Cote. 2009. "Record Densities of Indo-Pacific Lionfish on Bahamian Coral Reefs." *Coral Reefs* 28:107.

Gutierrez, Maria. 2007. "All That Is Air Turns Solid: The Creation of a Market for Carbon Sinks under the Kyoto Protocol. PhD diss., City University of New York.

Hall, Stuart. 1994. "Cultural Identity and Diaspora." In *Colonial Discourse and Postcolonial Theory,* edited by Patrick Williams and Laura Chrisman, 392–403. New York: Columbia University Press.
Hall-Campbell, Niambi. 2007. "Touring Poverty: Notes on Conducting Race and Inequality Fieldwork." *Community Psychologist* 40 (4): 55–56.
Hamilton, Clive. 2003. "The Commonwealth and Sea-Level Rise." *Round Table* 371 (1): 517–31.
Hanna, Liz, and Lachlan McIver. 2014. "Small Island States: Canaries in the Coal Mine of Climate Change and Health." In *Climate Change and Global Health,* edited by C.D. Butler, 181–92. Boston: CABI.
Haraway, Donna. 1989. "Introduction: The Persistence of Vision." In *Primate Visions: Gender, Race, and Nature in the World of Modern Science,* 1–15. Abingdon-on-Thames, UK: Routledge.
———. 2008. *When Species Meet.* Minneapolis: University of Minnesota Press.
———. 2016. *Staying with the Trouble: Making Kin in the Chthulucene.* Durham, NC: Duke University Press.
Hardin, Garrett. 1968. "The Tragedy of the Commons." *Science* 162:1243–48.
Hare, Lizzy. 2015. "The Anthropocene Trading Zone: The New Conservation, Big Data Ecology, and the Valuation of Nature." *Environment and Society: Advances in Research* 6:109–27.
Harrison, Mark. 2005. "Science and the British Empire." *History of Science Society* 96 (1): 56–63.
Hau'ofa, Epeli. 2008. *We Are the Ocean: Selected Works.* Honolulu: University of Hawai'i Press.
Hawkins, Ann. 1993. "Contested Ground: International Environmentalisms and Global Climate Change." In *The State and Social Power in Global Environmental Politics,* edited by Ronnie Lipschutz and Ken Conca, 221–45. New York: Columbia University Press.
Hawkins, Julie P., and Callum M. Roberts. 2004. "Effects of Artisanal Fishing on Caribbean Coral Reefs." *Conservation Biology* 18 (1): 215–26.
Hayden, Cori. 2003. *When Nature Goes Public: The Making and Unmaking of Bioprospecting in Mexico.* Princeton, NJ: Princeton University Press.
Head, Leslie. 2007. "Cultural Ecology: The Problematic Human and the Terms of Engagement." *Progress in Human Geography* 31 (6): 837–46.
Helmreich, Stefan. 2003. "Life@Sea: Networking Marine Biodiversity and Biotech Futures." In Franklin and Lock 2003, 227–59.
———. 2005. "How Scientists Think, about "Natives," for Example: A Problem of Taxonomy among Biologists of Alien Species in Hawaii." *Journal of the Royal Anthropological Institute* 11 (1): 107–28.
———. 2009. *Alien Ocean: Anthropological Voyages in Microbial Seas.* Berkeley: University of California Press.
Herskovits, Melville J. (1937) 1971. *Life in a Haitian Valley.* New York: Anchor Books.
Himes, Amber H. 2007. "Fishermen's Opinions of MPA Performance in the Egadi Islands Marine Reserve." *MAST* 5 (2): 55–76.
Holling, Crawford Stanley. 2001. "Understanding the Complexity of Economic, Ecological, and Social Systems." *Ecosystems* 4 (5): 390–405.

Holling, Crawford Stanley, Lance H. Gunderson, and Don Ludwig. 2002. "In Search of a Theory of Adaptive Change." In *Panarchy: Understanding Transformations in Systems of Humans and Nature,* edited by Lance Gunderson and Crawford Stanley Holling, 1–14. Washington, DC: Island.

Holm, Poul, Michael Evan Goodsite, Sierd Cloetingh, Mauro Agnoletti, Bedrich Moldan, Daniel J. Lang, Rik Leemans, et al. 2013. "Collaboration between the Natural, Social, and Human Sciences in Global Change Research." *Environmental Science and Policy* 28:25–35.

Howe, Cymene. 2014. "Anthropocene Ecoauthority: The Winds of Oaxaca." *Anthropological Quarterly* 87 (2): 381–404.

Howe, Cymene, and Anand Pandian. 2016. "Introduction: Lexicon for an Anthropocene Yet Unseen." *Cultural Anthropology,* January 21, 2016. https://culanth.org/fieldsights/788-introduction-lexicon-for-an-anthropocene-yet-unseen.

International Geosphere-Biosphere Programme. 2010. "Vision." Accessed February 26, 2019. www.igbp.net/about/vision.4.1b8ae20512db692f2a680001759 0.html.

Irvine, Richard, and Mina Gorji. 2013. "John Clare in the Anthropocene." *Cambridge Anthropology* 31 (1): 119–32.

Islands of The Bahamas. 2016. "About The Bahamas." Accessed February 26, 2019. www.bahamas.com/about-bahamas.

———. n.d. Official site. Accessed March 5, 2019. www.bahamas.com/.

Jackson, Jeremy B. C., Michael X. Kirby, Wolfgang H. Berger, Karen A. Bjorndal, Louis W. Botsford, Bruce J. Bourque, Roger H. Bradbury, et al. 2001. "Historical Overfishing and the Recent Collapse of Coastal Ecosystems." *Science* 293 (5530): 629–37.

Jefferson, Alan, and Leonard Lickorish. 1991. *Marketing Tourism: A Practical Guide.* Harlow, UK: Longman.

Julca, Alex, and Oliver Paddison. 2009. "Vulnerabilities and Migration in Small Island Developing States in the Context of Climate Change." *Natural Hazards* 55 (3): 717–28.

Keegan, William F. 1992. *The People Who Discovered Columbus: The Prehistory of The Bahamas.* Ripley P. Bullen Series. Gainesville: Florida Museum of Natural History.

Keen, Cathy. 2008. "Fossils from Bahamian Blue Hole May Give Clues to Early Life." May 1, 2008. www.floridamuseum.ufl.edu/science/fossils-from-bahamian-blue-hole-may-give-clues-to-early-life/.

Kelman, Ilan. 2010a. "Hearing Local Voices from Small Island Developing States for Climate Change." *Local Environment* 15 (7): 605–19.

———. 2010b. "Natural Disasters Do Not Exist (Natural Hazards Do Not Exist Either)." Version 3. July 9, 2010. www.ilankelman.org/miscellany/NaturalDisasters.doc.

Kelman, Ilan, and Jennifer West. 2009. "Climate Change and Small Island Developing States: A Critical Review." *Ecological and Environmental Anthropology* 5 (1): 1–16.

Kelty, Chris. 2006. *Collaborative Research: Agents of Mediation between Science and the Public*. NSF Proposal No. 0433457. Washington, DC: National Science Foundation.

Khan, Aisha. 2001. "Journey to the Center of the Earth: The Caribbean as Master Symbol." *Cultural Anthropology* 16 (3): 271–302.

———. 2004. *Callaloo Nation: Metaphors of Race and Religious Identity among South Asians in Trinidad*. Durham, NC: Duke University Press.

Kincaid, Jamaica. 1988. *A Small Place*. New York: Farrar, Straus and Giroux.

Kipnis, Andrew. 2008. "Audit Cultures: Neoliberal Governmentality, Socialist Legacy, or Technologies of Governing?" *American Ethnologist* 35 (2): 275–89.

Kirch, Patrick V. 1997. "Microcosmic Histories: Island Perspectives on 'Global' Change." *American Anthropologist* 99 (1): 30–42.

Kirksey, Eben. 2014. *The Multispecies Salon*. Durham, NC: Duke University Press.

Kirksey, S. Eben, and Stefan Helmreich. 2010. "The Emergence of Multispecies Ethnography." *Cultural Anthropology* 25 (4): 545–76.

Kloor, Keith. 2013. "Is the Anthropocene Doomed?" *Discover*, January 29, 2013. http://blogs.discovermagazine.com/collideascape/2013/01/29/is-the-anthropocene-doomed/#.UYhDqSsjppb.

Knorr Cetina, Karen. 1999. *Epistemic Cultures: How the Sciences Make Knowledge*. Cambridge, MA: Harvard University Press.

Knudson, Daniel C., Charles Greer, and Michelle Metro-Roland, eds. 2008. *Landscape, Tourism, and Meaning*. Farnham, UK: Ashgate.

Kohler, Robert E. 2002. *Landscapes and Labscapes: Exploring the Lab-Field Border in Biology*. Chicago: University of Chicago Press.

Kolbert, Elizabeth. 2014. *The Sixth Extinction: An Unnatural History*. New York: Holt.

Kress, John W., and Jeffrey K. Stine, eds. 2017. *Living in the Anthropocene: Earth in the Age of Humans*. Washington, DC: Smithsonian Books.

Krishna, Venni V. 1992. "The Colonial 'Model' and the Emergence of National Science in India: 1876–1920." In *Science and Empires*, edited by Patrick Petitjean, Catherine Jami, and Anne Marie Moulin, 57–72. Dordrecht: Kluwer Academic.

Kuklick, Henrika. 1996. "Islands in the Pacific: Darwinian Biogeography and British Anthropology. *American Ethnologist* 23 (3): 611–38.

Laarman, Jan G., and Richard Perdue. 1989. "Science Tourism in Costa Rica." *Annals of Tourism Research* 5 (16): 205–15.

LaFlamme, Alan G. 1979. "The Impact of Tourism: A Case from the Bahama Islands." *Annals of Tourism Research* 6 (2): 137–48.

Lahsen, Myanna. 2005. "Seductive Simulations? Uncertainty Distribution around Climate Models." *Social Studies of Science* 35:895–922.

———. 2010. "The Social Status of Climate Change Knowledge: An Editorial Essay." *WIRE: Climate Change* 1:162–71.

Latour, Bruno. 2004. *The Politics of Nature: How to Bring the Sciences into Democracy*. Cambridge, MA: Harvard University Press.

———. 2005. *Reassembling the Social: An Introduction to Actor Network Theory*. Oxford: Oxford University Press.

———. 2013. "Facing Gaia: Six Lectures on the Political Theology of Nature." *Figure/Ground Communication*, July 3, 2013. www.bruno-latour.fr/node/700.

———. 2014. "Anthropology at the Time of the Anthropocene: A Personal View of What Is to Be Studied." Distinguished Lecture, American Association of Anthropologists, Washington, DC, December 2014.

Lazrus, Heather. 2009. "The Governance of Vulnerability: Climate Change and Agency in Tuvalu, South Pacific." In Crate and Nuttall 2009, 240–49.

———. 2012. "Sea Change: Island Communities and Climate Change." *Annual Review of Anthropology* 41:285–301.

Lemelin, Harvey, Jackie Dawson, Emma J. Stewart, Pat Maher, and Michael Lueck. 2010. "Last-Chance Tourism: The Boom, Doom, and Gloom of Visiting Vanishing Destinations." *Current Issues in Tourism* 13 (5): 477–93.

Lemov, Rebecca. 2005. *World as Laboratory: Experiments with Mice, Mazes, and Men*. New York: Hill and Wang.

Lightbourn, Tiffany. 2000. "When Diasporas Discriminate: Identity Choices and Anti-immigrant Sentiment in The Bahamas." PhD diss., University of Michigan.

London, James. 2004. "Implications of Climate Change in Small Island Developing States: Experience in the Caribbean Region." *Journal of Environmental Planning and Management* 47 (4): 491–501.

Lorimer, Jamie. 2012. "Multinatural Geographies for the Anthropocene." *Progress in Human Geography* 36 (5): 1–20.

Lovbrand, Eva, Johannes Stripple, and Bo Wiman. 2009. "Earth System Governmentality: Reflections on Science in the Anthropocene." *Global Environmental Change* 19 (1): 7–13.

Lovelock, James E., and Lynn Margulis. 1974. "Atmospheric Homeostasis by and for the Biosphere: The Gaia Hypothesis." *Tellus* 26 (1-2): 2–10.

Lowe, Celia. 2006. *Wild Profusion: Biodiversity Conservation in an Indonesian Archipelago*. Princeton, NJ: Princeton University Press.

Luhmann, Niklas. 1993. "Deconstruction as Second-Order Observing." *New Literary History* 24 (4): 763–82.

———. 1998. *Observations on Modernity*. Translated by William Whobrey. Stanford, CA: Stanford University Press.

MacCannell, Dean. 1976. *The Tourist*. New York: Schocken.

Malinowski, Bronislaw. 1922. *Argonauts of the Western Pacific*. London: Macmillan.

Mallard, Grégoire, and Catherine Paradeise. 2009. "Global Science and National Sovereignty: A New Terrain for the Historical Sociology of Science." In Mallard, Paradeise, and Peerbaye 2009, 1–39.

Mallard, Grégoire, Catherine Paradeise, and Ashveen Peerbaye, eds. 2009. *Global Science and National Sovereignty: Studies in Historical Sociology of Science*. New York: Routledge.

Malm, Thomas. 2006. "No Island Is an 'Island': Some Perspectives on Human Ecology and Development in Oceania." In *The World and the Earth System: Global Socioenvironmental Change and Sustainability since the Neolithic,* edited by Alf Hornborg and Carole L. Crumley, 268–79. Walnut Creek, CA: Left Coast.

Marcus, George. 1995. "Ethnography in/of the World System: The Emergence of Multi-sited Ethnography." *Annual Review of Anthropology* 24:95–117.

———. 2002. "Intimate Strangers: The Dynamics of (Non) Relation between the Natural and Human Sciences." *Anthropological Quarterly* 75 (3): 519–26.

Martin, Romina, and Maja Schluter. 2015. "Combining System Dynamics and Agent-Based Modeling to Analyze Social-Ecological Interactions: An Example from Modeling Restoration of a Shallow Lake." *Frontiers in Environmental Science* 3 (October): 1–15.

Maurer, Bill. 1997. *Recharting the Caribbean: Land, Law, and Citizenship in the British Virgin Islands.* Ann Arbor: University of Michigan Press.

———. 2000. "A Fish Story: Rethinking Globalization on Virgin Gorda, British Virgin Islands." *American Ethnologist* 27 (3): 670–701.

McCarthy, James J., Osvaldo F. Canziani, Neil A. Leary, David J. Dokken, and Kasey S. White, eds. 2001. "Climate Change 2001: Impacts, Adaptation, and Vulnerability." In *Contribution of Working Group II to the Third Assessment Report of the Intergovernmental Panel on Climate Change,* 1–1005. Cambridge: Cambridge University Press.

McCay, Bonnie. 1978. "Systems Ecology, People Ecology, and the Anthropology of Fishing Communities." *Human Ecology* 6 (4): 397–422.

———. 2001. "Environmental Anthropology at Sea." In *New Directions in Anthropology and Environment,* edited by Carole L. Crumley, 257–72. Plymouth, UK: Altamira.

———. 2008. "The Littoral and the Liminal: Challenges to the Management of the Coastal and Marine Commons." *MAST* 7 (1): 7–30.

McCay, Bonnie J., and James Acheson, eds. 1987. *Question of the Commons: The Culture and Ecology of Communal Resources.* Tucson: University of Arizona Press.

McClanahan, Timothy, Jamie Davies, and Joseph Mania. 2005. "Factors Influencing Resource Users and Managers' Perceptions towards Marine Protected Area Management in Kenya. *Environmental Conservation* 32 (1): 42–49.

McConney, Patrick. 2002. "A Small Island Developing States (SIDS) and Social Science Perspective." In *FAO Fisheries Report No. 672: Report and Documentation of the International Work-Shop on Factors Contributing to Unsustainability and Overexploitation in Fisheries,* edited by Dominique Greboval, 151–54. Rome: Food and Agriculture Organization of the United Nations.

McNeill, John Robert, and Peter Engelke. 2014. *The Great Acceleration: An Environmental History of the Anthropocene since 1945.* Cambridge, MA: Belknap Press of Harvard University Press.

Meadows, Dennis L., Donella H. Meadows, Jorgen Randers, and William W. Behrens. 1972. "The Limits to Growth, London." Washington, DC: Potomac Associates.

Meyer, Robinson. 2018. "Geologies Time Keepers Are Feuding." *Atlantic,* July 20, 2018. www.theatlantic.com/science/archive/2018/07/anthropocene-holocene-geology-drama/565628/.

Michener, William K., Thomas J. Baerwald, Penelope Firth, Margaret A. Palmer, James L. Rosenberger, Elizabeth A. Sandlin, and Herman Zimmerman. 2001. "Defining and Unraveling Biocomplexity." *Bioscience* 51 (2): 1018–23.

Mimura, Nobuo, Leonard Nurse, Roger F. McLean, John Agard, Lino Briguglio, Penehuro Lefale, Rolph Payet, et al. 2007. "Small Islands." In Parry et al. 2007, 687–716.

Mintz, Sidney. 1983. *Sweetness and Power: The Place of Sugar in Modern History.* London: Penguin Books.

Mol, Annemarie, and John Law. 2002. "Complexities: An Introduction." *Complexities: Social Studies of Knowledge Practices.* Durham, NC: Duke University Press.

Monastersky, Richard. 2015. "Anthropocene: The Human Age." *Nature* 519 (7542): 144–47.

Moore, Amelia. 2010. "Climate Changing Small Islands: Considering Social Science and the Production of Island Vulnerability and Opportunity." *Environment and Society: Advances in Research* 1 (1): 116–31.

———. 2012. "The Aquatic Invaders: Marine Management Figuring Fishermen, Fisheries, and Lionfish." *Cultural Anthropology* 24 (7): 667–88.

———. 2015a. "The Anthropocene: A Critical Exploration." *Environment and Society: Advances in Research* 6:1–3.

———. 2015b. "Islands of Difference: Design, Urbanism, and Sustainable Tourism in the Anthropocene Caribbean." *Journal of Latin American and Caribbean Anthropology* 20 (3): 513–32.

———. 2015c. "Tourism in the Anthropocene Park? New Analytic Possibilities." *International Journal of Tourism Anthropology* 4 (2): 186–200.

———. 2016. "Anthropocene Anthropology: Reconceptualizing Contemporary Global Change." *Journal of the Royal Anthropological Institute* 22 (1): 27–46.

Moore, Donald S. 2000. "The Crucible of Cultural Politics: Reworking Development in Zimbabwe's Eastern Highlands." *American Ethnologist* 26 (3): 654–89.

Moore, Donald, Jake Kosek, and Anand Pandian, eds. 2003. *Race, Nature, and the Politics of Difference.* Durham, NC: Duke University Press.

Moore, George W. 1964. *Speleology: The Study of Caves.* Lexington, MA: Heath.

Moore, Jason W. 2016. *Anthropocene or Capitaloscene? Nature, History, and the Crisis of Capitalism.* Oakland, CA: PM Press.

Moore, Rawleston. 2002. "Vulnerability and Adaptation: A Regional Synthesis of the Vulnerability and Adaptation Components of Caribbean National Communications." *Caribbean Planning for Adaptation to Global Climate Change Project.* Accessed February 26, 2019. https://research.fit.edu/media/site-specific/researchfitedu/coast-climate-adaptation-library/latin-america-and-caribbean/regional---caribbean/CPACC.--2002.--Caribbean-Vulnerabililty--Adaptation-to-CC..pdf.

Mora, Camilo, and Peter F. Sale. 2011. "Ongoing Global Biodiversity Loss and the Need to Move beyond Protected Areas: A Review of the Technical and

Practical Shortcomings of Protected Areas on Land and Sea." *Marine Ecology Press Series* 434:251–66.
Morris, James A., Jr., and John L. Akins. 2008. "Feeding Ecology of Invasive Lionfish *(Pterois volatins)* in the Bahamian Archipelago." *Environmental Biology of Fishes* 86:389–98.
Moss, Richard H., Jae A. Edmonds, Kathy A. Hibbard, Martin R. Manning, Steven K. Rose, Detlef P. van Vuuren, Timothy R. Carter, et al. 2010. "The Next Generation of Scenarios for Climate Change Research and Assessment." *Nature* 463:747–56.
National Science Foundation. 2015. "National Science Foundation Awards $9.47 Million for Research on Coupled Natural and Human Systems." Accessed February 26, 2019. www.nsf.gov/news/news_summ.jsp?cntn_id=132412.
Nature Conservancy. 2016. "The Bahamas." Accessed February 26, 2019. www.nature.org/ourinitiatives/regions/caribbean/bahamas/index.htm.
———. 2017. "Caribbean Challenge Initiative." Accessed February 26, 2019. www.nature.org/ourinitiatives/regions/caribbean/caribbean-challenge.xml.
Nelson, Donald R., Colin T. West, and Timothy J. Finan. 2009. "Introduction to 'In Focus: Global Change and Adaptation in Local Places.'" *American Anthropologist* 111 (3): 271–74.
Newsome, David, and Ross Dowling. 2010. *Geotourism: The Tourism of Geology and Landscape.* Oxford: Goodfellow.
Nixon, Angelique. 2015. *Resisting Paradise: Tourism, Diaspora, and Sexuality in Caribbean Culture.* Jackson: University Press of Mississippi.
Norum, Roger. 2013. "The Anthropocene Arctic." *Arctic Encounters Researcher's Blog.* 2017. www.arcticencounters.net/post.php?s = 2013–12–21-the-anthropocene-arctic (site discontinued).
NOVA. 2010. *Extreme Cave Diving.* PBS. February 9, 2010. www.pbs.org/wgbh/nova/cavedive/about.html.
Nowotny, Scott Gibbons. 2001. *Re-thinking Science: Knowledge and the Public in the Age of Uncertainty.* Malden, MA: Blackwell
Ogden, Laura, Nik Heynen, Ulrish Oslender, Paige West, Karim Aly-Kassam, and Paul Robbins. 2013. "Global Assemblages, Resilience, and Earth Stewardship in the Anthropocene." *Frontiers in Ecology and the Environment* 11 (7): 341–47.
Oliver-Smith, Anthony. 2009. "Climate Change and Population Displacement: Disasters and Diasporas in the Twenty-First Century." In Crate and Nuttall 2009, 116–36.
Olson, Valerie. 2010. "The Ecobiopolitics of Space Biomedicine." *Medical Anthropology* 29 (2): 170–93.
Olson, Valerie, and Lisa Messeri. 2015. "Beyond the Anthropocene: Un-earthing an Epoch." *Environment and Society: Advances in Research* 6:28–47.
Ong, Aihwa, and Stephen J. Collier, eds. 2005. *Global Assemblages: Technology, Politics, and Ethics as Anthropological Problems.* Malden, MA: Blackwell.
Ortiz, Fernando. 1947. *Cuban Counterpoint: Tobacco and Sugar.* New York: Knopf.

Osborne, Michael, Miles Traer, and Leslie Chang. 2013. *Generation Anthropocene: Stories and Conversations about Planetary Change.* Accessed February 26, 2019. www.genanthro.com/.

Paddack, Michelle J., John D. Reynolds, Consuelo Aguilar, Richard S. Appeldoorn, Jim Beets, Edward W. Burkett, Paul M. Chittaro, et al. 2009. "Recent Region-Wide Declines in Caribbean Reef Fish Abundance." *Current Biology* 19:590–95.

Paley, Julia. 2001. "Making Democracy Count: Opinion Polls and Market Surveys in Chilean Political Transition." *Cultural Anthropology* 16 (2): 135–64.

Palsson, Gisli. 1991. *Coastal Economies, Cultural Accounts: Human Ecology and Icelandic Discourse.* Manchester: Manchester University Press.

———. 1994. "Enskilment at Sea." *Man* 29 (4): 901–27.

Palsson, Gisli, and E. Paul Durrenberger. 1990. "Systems of Production and Social Discourse: The Skipper Effect Revisited." *American Anthropologist* 92 (1): 130–41.

Parry, Martin L., Osvaldo F. Canziani, Jean P. Palutikof, Pierre J. van der Linden, and C. E. Hanson, eds. 2007. *Climate Change 2007: Impacts, Adaptation and Vulnerability; Contribution of Working Group II to the Fourth Assessment Report of the Intergovernmental Panel on Climate Change.* Cambridge: Cambridge University Press.

Patz, Jonathan A., Diarmid Campbell-Lendrum, Tracey Holloway, and Jonathan A. Foley. 2005. "Impact of Regional Climate Change on Human Health." *Nature* 438:310–17.

Pauly, Daniel. 2006. "Major Trends in Small-Scale Marine Fisheries, with Emphasis on Developing Countries, and Some Implications for the Social Sciences." *MAST* 4 (2): 7–22.

Pelling, Mark, and Juha I. Uitto. 2001. "Small Island Developing States: Natural Disaster Vulnerability and Global Change." *Global Environmental Change Part B: Environmental Hazards* 3 (2): 49–62.

Peterson, Nicole, and Kenneth Broad. 2009. "Climate and Weather Discourse in Anthropology: From Determinism to Uncertain Futures." In Crate and Nuttall 2009, 70–86.

Pilgrim, Sophie. 2014. "Video: Sinking Islands 'Canary in Coal Mine' of Global Warming." *France* 24. October 18, 2014. www.france24.com/en/20141017-kiribati-sinking-islands-canary-coal-mine-global-warming.

Pomeroy, Robert S., John E. Parks, and Lani M. Watson. 2004. *How Is Your MPA Doing? A Guidebook of Natural and Social Indicators for Evaluating Marine Protected Area Management Effectiveness.* Gland, Switzerland: International Union for Conservation of Nature.

Post, James. 2008. *Carbon Free Vacation.* Informational pamphlet advertisement. Grenada: Paradise Bay Resort.

Poucher, Sandra, and Rick Copeland. 2006. *Speleological and Karst Glossary of Florida and the Caribbean.* Gainesville: University Press of Florida.

Pugh, Jonathan, David Chandler, and Elaine Stratford. 2015. *Islands, Archipelagos, and the Anthropocene (2): Contemporary Debates in Island Studies.* Panel Presented at the Royal Geographical Society Annual Conference,

Exeter, UK. Accessed February 26, 2019. http://conference.rgs.org/AC2015/247.
Rabinow, Paul. 1989. *French Modern: Norms and Forms of the Social Environment.* Cambridge, MA: MIT Press.
———. 1996. "Artificiality and Enlightenment: From Sociobiology to Biosociality." In *Essays on the Anthropology of Reason,* 91–111. Princeton, NJ: Princeton University Press.
———. 1999. *French DNA: Trouble in Purgatory.* Chicago: University of Chicago Press.
———. 2003. *Anthropos Today: Reflections on Modern Equipment.* Princeton, NJ: Princeton University Press.
———. 2008. *Marking Time: On the Anthropology of the Contemporary.* Princeton, NJ: Princeton University Press.
Rabinow, Paul, and Nicolas Rose. 1994. *The Essential Foucault.* New York: New Press.
Radcliffe-Brown, Alfred Reginald. 1922. *The Andaman Islanders.* New York: Free Press of Glencoe.
Raffles, Hugh. 2002. *In Amazonia: A Natural History.* Princeton, NJ: Princeton University Press.
Richardson, Sarah S., and Hallam Stevens, eds. 2015. *Postgenomics: Perspectives on Biology after the Genome.* Durham, NC: Duke University Press.
Robbins, Paul, and Sarah. A. Moore. 2013. "Ecological Anxiety Disorder: Diagnosing the Politics of the Anthropocene." *Cultural Geographies* 20 (1): 3–19.
Rockstrom, Johan. 2009. "A Safe Operating Space for Humanity." *Nature* 461:472–75.
Rodman, Margaret C. 1987. "Constraining Capitalism? Contradictions in Self-Reliance in Vanuatu Fisheries Development." *American Ethnologist* 14 (4): 712–26.
Roncoli, Carla, Todd Crane, and Ben Orlove. 2009. "Fielding Climate Change in Cultural Anthropology." In Crate and Nuttall 2009, 87–115.
Rose, Nikolas. 2006. *The Politics of Life Itself: Biomedicine, Power, and Subjectivity on the 21st Century.* Princeton, NJ: Princeton University Press.
Rose, Nikolas, and Carlos Novas. 2005. "Biological Citizenship." In Ong and Collier 2005, 439–63. Malden, MA: Blackwell.
Rosenthal, Elisabeth. 2011. "Answer for Invasive Species: Put It on a Plate and Eat It." *New York Times,* July 9, 2011, A14.
Rounsevell, Mark D.A., Derek T. Robinson, and Dave Murray-Rust. 2012. "From Actors to Agents in Socio-ecological Systems Models." *Philosophical Transactions of the Royal Society B* 367 (1586): 259–69.
Rubenstein, Daniel I., and Richard W. Wrangham, eds. 1987. *Ecological Aspects of Social Evolution: Birds and Mammals.* Princeton, NJ: Princeton University Press.
Rudiak-Gould, Paul. 2010. "Promiscuous Corroboration and Climate Change Translation: A Case Study from the Marshall Islands." *Global Environmental Change* 22 (1): 46–54.

———. 2013. "We Have Seen It with Our Own Eyes: Why We Disagree about Climate Change Visibility." *Weather, Climate, and Society* 5 (2): 120–32.

Saunders, Gail, and Michael Craton. 1992. "The End of the Old Regime, 1763–1783." In *From Aboriginal Times to the End of Slavery*, 157–78. Vol. 1 of *Islanders in the Stream: A History of the Bahamian People*. Athens: University of Georgia Press.

———. 1998. *From the Ending of Slavery to the Twenty-First Century*. Vol. 2 of *Islanders in the Stream: A History of the Bahamian People*. Athens: University of Georgia Press.

Sayre, Nathan. 2012. "The Politics of the Anthropogenic." *Annual Review of Anthropology* 41:57–70.

Schiermeier, Quirin. 2004. "Climate Findings Let Fishermen Off the Hook." *Nature* 428 (6978): 4.

Schneider, Stephen H., S. Semenov, Anand Patwardhan, Ian Burton, Chris H.D. Magadza, Michael Oppenheimer, A. Barrie Pittock, et al. 2007. "Assessing Key Vulnerabilities and the Risk from Climate Change." In Parry et al. 2007, 779–810.

Schwabe, Stephanie, James Carew, and Rob Palmer. 2006. "Blue Holes: An Inappropriate Moniker for Scientific Discussion of Water-Filled Caves in The Bahamas." In *12th Symposium on the Geology of The Bahamas and Other Carbonate Regions*. Accessed February 26, 2019. https://blueholes.org/The%20Rob%20Palmer%20Blue%20Holes%20Foundation/Publications_files/12th%20Inappropriate%20name.pdf, 1–9.

———. 2007. "Making Caves in The Bahamas: Different Recipes, Same Ingredients." In *13th Symposium on the Geology of The Bahamas and other Carbon Regions*. www.blueholes.org/The%20Rob%20Palmer%20Blue%20Holes%20Foundation/Publications_files/13.%20Schwabe%20Making-Caves%20Final%204.13.08.pdf, 153–68.

Scoones, Ian. 1999. "New Ecology and the Social Sciences: What Prospects for a Fruitful Engagement?" *Annual Review of Anthropology* 28: 479–507.

Scott, David. 2004. *Conscripts of Modernity: The Tragedy of Colonial Enlightenment*. Durham, NC: Duke University Press.

Scranton, Roy. 2015. *Learning to Die in the Anthropocene: Reflections on the End of a Civilization*. San Francisco: City Lights Open Media.

Segal, Daniel A. 2001. "Editor's Note: Anthropology and/in/of Science." *Cultural Anthropology* 16 (4): 451–52.

Seidl, Roman, Fridolin Simon Brand, Michael Stauffacher, Pius Krütli, Quang Bao Le, Andy Spörri, Grégoire Meylan, et al. 2013. "Science with Society in the Anthropocene." *AMBIO* 42:5–12.

Sheller, Mimi. 2003. *Consuming the Caribbean: From Arawaks to Zombies*. Abington-on-Thames, UK: Routledge.

Singleton, Benedict. 2012. "Anthropocene Nights." *Architectural Design* 82 (4): 66–71.

Slaughter, Richard A. 2012. "Welcome to the Anthropocene." *Futures* 44 (2): 119–26.

Smajgl, Alex, Daniel G. Brown, Diego Valbuena, and Marco G.A. Huigen. 2011. "Empirical Characterization of Agent Behaviors in Socio-ecological Systems." *Environmental Modeling and Software* 26:837–44.

Smith, M. Estellie, ed. 1977. *Those Who Live from the Sea: A Study in Maritime Anthropology.* New York: West.

Smith, Michael Garfield. 1965. *The Plural Society in the British West Indies.* Berkeley: University of California Press.

Smith, Valene, ed. 1989. *Hosts and Guests: The Anthropology of Tourism.* Philadelphia: University of Pennsylvania Press.

Sommer, Gunilla, and James Carrier. 2010. "Tourism and Its Others: Tourists, Traders and Fishers in Jamaica." In *Tourism, Power and Culture: Anthropological Insights,* edited by Donald Macleod and James Carrier, 174–96. Bristol, UK: Channel View.

Spoehr, Alexander, ed. 1980. *Maritime Adaptations: Essays on Contemporary Fishing Communities.* Pittsburgh: University of Pittsburgh Press.

Stengers, Isabelle. 2010. "Including Nonhumans in Political Theory: Opening Pandora's Box?" In Braun and Whatmore 2010, 3–34.

Steward, Julian H. 1955. *Theory of Culture Change: The Methodology of Multilinear Evolution.* Champaign: University of Illinois Press.

Stoffle, Rich, and Jessica Minnis. 2007. "Marine Protected Areas and the Coral Reefs of Traditional Settlements in the Exumas, Bahamas." *Coral Reefs* 26:1023–32.

Strachan, Ian. 2002. *Paradise and Plantation: Tourism and Culture in the Anglophone Caribbean.* Charlottesville: University of Virginia Press.

Strang, Veronica. 2009. "Integrating the Social and Natural Sciences in Environmental Research: A Discussion Paper." *Environment Development Sustainability* 11:1–18.

Strathern, Marilyn. 2005. "Robust Knowledges and Fragile Futures." In Ong and Collier 2005, 464–81.

"*Strombus gigas.*" n.d. Accessed February 26, 2019. https://cites.org/eng/gallery/species/invertibrate/queen_conch.html, app. 2.

Stronza, Amanda. 2001. "Anthropology of Tourism: Forging New Ground for Ecotourism and Other Alternatives." *Annual Review of Anthropology* 30:261–83.

Subcommission on Quaternary Stratigraphy. 2015. "Working Group on the 'Anthropocene.'" Accessed February 26, 2019. http://quaternary.stratigraphy.org/workinggroups/anthropocene/.

Subramaniam, Banu. 2001. "The Aliens Have Landed! Reflections on the Rhetoric of Biological Invasions." *Meridians: Feminism, Race, Transnationalism* 2 (1): 26–40.

Sundaravadanan, Arathi. 2009. "The Bahamas: Stopping the Lionfish." Nature Conservancy. Accessed 2012. www.nature.org/wherewework/caribbean/bahamas/features/ (site discontinued).

Sunder Rajan, Kaushik. 2006. *Biocapital: The Constitution of Postgenomic Life.* Durham, NC: Duke University Press.

Swanson, Heather, Nils Bubandt, and Anna Tsing. 2015. "Less Than One but More Than Many: Anthropocene as Science Fiction and Scholarship-in-the-Making." *Environment and Society: Advances in Research* 6:149–66.

Takacs, David. 1996. *The Idea of Biodiversity: Philosophies of Paradise*. Baltimore: Johns Hopkins University Press.

Teaiwa, Katerina Martina. 2014. *Consuming Ocean Island: Stories of People and Phosphate from Banaba*. Bloomington: Indiana University Press.

Thomas, Deborah A. 2004. *Modern Blackness: Nationalism, Globalization, and the Politics of Culture in Jamaica*. Durham, NC: Duke University Press.

Thompson, Charis. 2010. "Asian Regeneration? Nationalism and Internationalism in Stem Cell Research." *Asian Biotech*, edited by Aihwa Ong and Nancy Chen, 95–117. Durham, NC: Duke University Press.

Thompson, Krista A. 2006. *An Eye for the Tropics: Tourism, Photography, and Framing the Caribbean Picturesque*. Durham, NC: Duke University Press.

Trouillot, Michel-Rolph. 1992. "The Caribbean Region: An Open Frontier in Anthropological Theory." *Annual Review of Anthropology* 21:19–42.

Tsing, Anna. 2005. *Friction: An Ethnography of Global Connection*. Princeton, NJ: Princeton University Press.

UNESCO World Heritage Center. 2014. "The World Heritage Convention Has Entered into Force for The Bahamas." August 15, 2014. http://whc.unesco.org/en/news/1177/.

United Nations. 2005a. *Caribbean Environmental Outlook: Special Edition for the Mauritius International Meeting for the 10 Year Review of the Barbados Programme of Action*. Tokyo: United Nations Press.

———. 2005b. *Designing Household Survey Samples: Practical Guidelines*. Studies in Methods. Series F(98). New York: United Nations Statistics Division.

UNWTO (World Tourism Organization). 2008. *Climate Change and Tourism: Responding to Global Challenges*. Madrid: WTO/United Nations Environment Programme.

Van der Velde, Marijn, Steven R. Green, Marnik Vanclooster, and Brent E. Clothier. 2007. "Sustainable Development in Small Island Developing States: Agricultural Intensification, Economic Development, and Fresh-Water Resources Management on the Coral Atoll of Tongatapu." *Ecological Economics* 61 (2–3): 456–68.

Van Ginkel, Rob, and Jojada Verrips. 2002. "Maritime Studies (MAST): An Editorial Reintroduction." *MAST* 1 (1): 5–7.

Wackernagel, Mathis, Niels B. Schulz, Diana Deumling, Alejandro Callejas Linares, Martin Jenkins, Valerie Kapos, Chad Monfreda, et al. 2002. "Tracking the Ecological Overshoot of the Human Economy." *Proceedings of the National Academy of Sciences* 99 (14): 9266–71.

Waters, Colin, and Jan Zalasiewicz. 2013. "Report of Activities 2012." *Newsletter of the Anthropocene Working Group* 4:1–18.

West, Paige. 2006. *Conservation Is Our Government Now: The Politics of Ecology on Papua New Guinea*. Durham, NC: Duke University Press.

———. 2008. "Tourism as Science and Science as Tourism: Environment, Society, Self, and Other in Papua New Guinea." *Current Anthropology* 49 (4): 597–626.

West, Paige, and James Carrier. 2004. "Ecotourism and Authenticity: Getting Away from It All?" *Current Anthropology* 45 (4): 483–98.

Whitty, Julia. 2009. "Listen to the Lionfish: What Invasive Species Are Trying to Tell Us." *Mother Jones* 34 (1): 60–65.

Wilson, Edward O., and Robert H. MacArthur. 1967. *The Theory of Island Biogeography*. Princeton, NJ: Princeton University Press.

Wise, Sarah, and Amelia Moore. 2013. "Underground, Undersea: The Uses and Perceptions of Bahamian Blue Holes." *Underwater Speleology* 40 (1): 20–23.

Woody Foundation. 2014. "4th Annual Lionfish Bash." Poster. Accessed January 23, 2019. www.woodyfoundation.org/event/4th-annual-lionfish-bash/.

World Travel and Tourism Council. 2017. *Bahamas*. Accessed February 26, 2019. www.wttc.org/-/media/files/reports/economic-impact-research/countries-2017/bahamas2017.pdf.

World Wildlife Fund. 2016. "Caribbean Islands: Bahamas." Accessed February 26, 2019. www.worldwildlife.org/ecoregions/nt0203.

Yearley, Steven. 2009. "Sociology and Climate Change after Kyoto: What Roles for Social Science in Understanding Climate Change?" *Current Sociology* 57 (3): 389–405.

Young, Louise B. 1999. *Islands: Portraits of Miniature Worlds*. New York: Freeman.

Young, Virginia Heyer. 1993. *Becoming West Indian: Culture, Self, and Nation in St. Vincent*. Washington, DC: Smithsonian Institution Press.

Zalasiewicz, Jan, Mark Williams, Will Steffen, and Paul Crutzen. 2010. "The New World of the Anthropocene." *Environmental Science and Technology* 44:2228–31.

Index